The Fractal Cosmos II

A DNA Analogue

Stephen Blaha Ph. D.
Blaha Research

MMXXV

.

ISBN: 979-8-9986071-5-8

Rev. 00/00/01 September 6, 2025

To Margaret

Some Other Books by Stephen Blaha

SuperCivilizations: Civilizations as Superorganisms (McMann-Fisher Publishing, Auburn, NH, 2010)

Unification of God Theory and Unified SuperStandard Model THIRD EDITION (Pingree Hill Publishing, Auburn, NH, 2018).

The Exact QED Calculation of the Fine Structure Constant Implies ALL 4D Universes have the Same Physics/Life Prospects (Pingree Hill Publishing, Auburn, NH, 2019).

Passing Through Nature to Eternity ProtoCosmos, HyperCosmos, Unified SuperStandard Theory (Pingree Hill Publishing, Auburn, NH, 2022).

HyperCosmos Fractionation and Fundamental Reference Frame Based Unification: Particle Inner Space Basis of Parton and Dual Resonance Models (Pingree Hill Publishing, Auburn, NH, 2022).

The Cosmic Panorama: ProtoCosmos, HyperCosmos, Unified SuperStandard Theory (UST) Derivation (Pingree Hill Publishing, Auburn, NH, 2022).

God and and Cosmos Theory (Pingree Hill Publishing, Auburn, NH, 2023).

Newton's Apple is Now The Fermion (Pingree Hill Publishing, Auburn, NH, 2023).

Cosmos Theory: The Sub-Particle Gambol Model (Pingree Hill Publishing, Auburn, NH, 2023).

Cosmos-Universe-Particle-Gambol Theory (Pingree Hill Publishing, Auburn, NH, 2024).

Structure and Dynamics of Cosmos Theory and the Unified SuperStandard Theory (Pingree Hill Publishing, Auburn, NH, 2024).

Black Holes, White Holes, and Superluminal Starship (Pingree Hill Publishing, Auburn, NH, 2025).

The Sequence Theory Foundation of Cosmos Theory and Gambol Theory (Pingree Hill Publishing, Auburn, NH, 2025).

BEYOND TIME AND SPACE Fermion Strata & Consistency Conditions for Sequences, Cosmos Spaces and Gambols (Pingree Hill Publishing, Auburn, NH, 2025).

The C o s m o s as a Physics Organism (Blaha Research, Auburn, NH, 2025)

The Root of Reality (Blaha Research, Auburn, NH, 2025)

Available on Amazon.com, bn.com, Amazon.co.uk and other international web sites as well as at better bookstores.

CONTENTS

FIGURES and TABLES

Introduction

This volume extends volume 1 of The Fractal Cosmos to provide new insights and extended detail on the fractal features and dynamics of Cosmos Theory and its fundamental particles and symmetries. The remarkable features of Cosmos Theory are further enhanced. The basis is a new representation of Cosmos Theory using fractals. It appears that the solidity of the Cosmos is striated into fractal components. Our view of Nature must now adapt to a somewhat striated entity composed of the strands of Physical Reality.

This book has a number of new features:

1. New definitions of fractal multiplication and division. Fractal multiplication is now commutative.

2. A two strand Cosmos Curve which upon further investigation becomes a compressed double helix that is analogous, in structure, to organic DNA.

3. A quantum probabilistic interpretation of universe creation with amplitudes for a universe with a Dark part, a universe without a Dark part, and a universe with a quantum superposition of a universe with a Dark part and a universe without a Dark part. This superposed universe could have a Dark part whose effect is only seen by gravitation.

4. A new view of fractal interaction coupling constants based on fractal SU(n) symmetries which now have dimension arrays – just like Cosmos spaces.

5. A new representation of fundamental fermion (lepton and quark) dynamics based on a new fractal form of Planckian Distribution.

6. A new representation of the fractal generation of universes based on a Lebesgue Z-order curve analysis.

7. A new sequence of massive gauge bosons for ElectroWeak Theory.

This book greatly extends the understanding of Cosmos Theory to a new deeper fractal formulation. All components are fractal! Nature is striated. Cosmos Theory connects the deepest parts of the Cosmos to experimental data such as masses and interactions. The study of the Cosmos continues.

1. Double Stranded Fractal Cosmos Curve

This chapter develops fractal arithmetic based on new forms of fractal multiplication and division. It then proceeds to develop more detail on the Fractal Cosmos Curve that leads to a (flattened) double helix Cosmos Curve.

1.1 Brief View of Fractals

The dimension D of simple and fractal curves is called the Hausdorff dimension. It can be an integer, as it is for simple spaces such as 3-dimension space where D = 3. It can be a non-integer value for a space filling curve such as the Hilbert curve where the line occupies a two dimension space.

1.1.1 Hausdorff Dimension

The Hausdorff dimension D of a curve is

$$D = \log(N)/\log(\varepsilon) = \log_\varepsilon(N) \tag{1.0}$$

where the numerator = log N equals the logarithm of the number of copies of the shape formed from each copy of the previous iteration of the curve under construction. The denominator = log ε is the logarithm of the factor by which copies are scaled downward from the previous iteration.

For example, a Sierpinski triangle (double sided) creates 3 copies resulting in N = 3 copies from the previous iteration. The scale is ε = 2 based on the doubling of the side. As a result:

$$D = \log(3)/\log(2) = \log_2(3) = 1.585$$

or

$$2^D = 3$$

For the Cosmos Curve and the Hilbert curve we find N = 4 and ε = 2 with the result:

$$D = \log(4)/\log(2) = 2$$

1.1.2 Fractal Cosmos Spaces and Fractal Dimension Arithmetic

The Cosmos Fractal Curve is described (created) in a series of steps – *conceptually*. In Reality it exists instantaneously. There is no point in time for its creation since time does not exist "before" the Cosmos Fractal Curve. So the "creation" process that we describe has the sole purpose of providing an understanding of the Cosmos. The Cosmos Fractal Curve fills a two dimension unit square. It has fractal dimension 2.

At each step of the Cosmos Fractal Curve construction there is a *numerical* value 2^n – the length of the piecewise linear curve used in construction. The number denoted n has only been used as the linear curve length until now. We will now expand its definition based on its value being related to the number of components (dimensions) in a Cosmos space. A HyperCosmos space of dimension r has a dimension array with 2^{r+4} components with n = r + 4. A 2nd Kind HyperCosmos space has a dimension array with 2^{r+3} components with n = r + 3.

We will extend the nature of the piecewise linear curve number n to have an associated Cosmos space with a dimension array size of $2^n = 2^{r+4}$. The definition of fractal dimension (including topological dimensions) may be applied to Cosmos spaces by setting N in eq. 1.0 with

$$N = 2^n$$

and ε = 2 with the result that

$$D(n) = \log(2^n)/\log(2) = \log_2(2^n) = n \tag{1.0a}$$

is the Fractal (topological) dimension of the associated space.

Each Cosmos space now has an associated fractal of fractal dimension $D(2^{r+4}) = r + 4$ (HyperCosmos spaces) or $D(r^{r+3}) = r + 3$ (2nd Kind HyperCosmos spaces). We are justified in associating construction step n with the Cosmos space of dimension n since there is a 1:1 map between construction steps and Cosmos spaces.

In addition to associating a HyperCosmos space with each construction step we will also associate a space-time[1] of dimension r = n – 4 at construction step n.[2] These space-times do not exist until universes are created. They become the space-times of the universes. As an associate of a construction step a space-time is an artifact of Cosmos structure and thus not Physically instantiated. See Fig. 1.6 for a diagram of a section of the Cosmos Fractal Curve together with its associated spaces.

At each construction step there is the length of the piecewise linear curve, a Cosmos space and a space-time.

We will find it necessary to "add", "subtract", "multiply", and "divide" fractals. When we perform arithmetic operations they will be applied to fractal dimensions although they will be exemplified with Cosmos dimension arrays. We define fractal dimensions from Cosmos dimension array sizes. Since the fractal dimensions are based on logarithms as in eq. 1.1 below the arithmetic operations on fractal dimensions will be based on logarithms of dimension array sizes. Thus, for example, fractal dimension addition is based on products of dimension array sizes. See eq. 1.3 below.

Fractal dimension arithmetic (in these discussions) is based on dimension array sizes and the nature of logarithms, which lead to a different form for arithmetic operations.

---

[11] For Cosmos space of dimension r there is a subset of r dimensions in its dimension array for its space-time.
[2] For 2nd Kind HyperCosmos spaces r = n – 3.

1.1.2.1 Fractal Addition

The addition of fractals does not seem to have been considered before volume 1. Later we will see fractal addition is implied in certain fractal relations between particles. We will study fractal addition in the case of Cosmos spaces.

There are several definitions of dimensions for Cosmos spaces:

1. We have assigned an even integer dimension r to each Cosmos space.
2. Each HyperCosmos space of dimension r has a dimension $D(2^{r+4})$ equal to the number of elements in its dimension array $d_{dr} = 2^{r+4}$.
3. Each 2nd Kind HyperCosmos space of dimension r has a dimension $D(2^{r+3})$ equal to the number of elements in its dimension array $d_{dr} = 2^{r+3}$.
4. Each HyperCosmos space and 2nd Kind HyperCosmos space of dimension r has a space-time dimension $D(2^{r+4})$ or $D(2^{r+3})$.

We define the fractal dimension of a HyperCosmos space of dimension r + 4 with

$$D(2^{r+4}) = \log(2^{r+4})/\log(2) = \log_2(2^{r+4}) = r + 4 \quad \text{for HyperCosmos Spaces} \qquad (1.1)$$

and for 2nd Kind HyperCosmos spaces as

$$D(2^{r+3}) = \log(2^{r+3})/\log(2) = \log_2(2^{r+3}) = r + 3 \quad \text{for 2nd Kind HyperCosmos Spaces}$$

See Figs. 1.1, 1.2, and 1.3. In general we define

$$D(2^n) = \log(2^n)/\log(2) = \log_2(2^n) = n \qquad (1.2)$$

and an inverse

$$D^{-1}(n) = 2^n$$
$$D^{-1}(D(2^n)) = 2^n$$
$$D(D^{-1}(n)) = n$$

We identify the dimension array sizes as corresponding to fractal dimensions. (They are also the lengths of piecewise linear line segments in the construction of the Cosmos fractal curve.)[3]

If we "add" two spaces of dimensions r and r′, the combined space has the dimension array size:

HyperCosmos Space: $d_{dr}\, d_{dr'} = 2^{r+4} * 2^{r'+4} = 2^{r+r'+8} = 16 d_{d(r+r')}$ (1.3)

2nd Kind HyperCosmos Space: $d_{dr2}\, d_{dr'2} = 2^{r+3} * 2^{r'+3} = 2^{r+r'+6} = 8 d_{d(r+r')2}$

The corresponding Hausdorff dimension "additions" give the combined dimensions as[4]

---

[33] These values change to 2^{r+7} and 2^{r+6} respectively when strata and fermion gambols are introduced. The results are similar with these choices.

$$D(2^{r+4}) \text{ “+” } D(2^{r'+4}) = D(2^{r+r'+8})$$
$$D(2^{r+3}) \text{ “+” } D(2^{r'+3}) = D(2^{r+r'+6}) \tag{1.4}$$

It is clear that dimension addition adds the powers of 2. We can interpret the “addition” in eq. 1.4 with blocks of dimension arrays as:

Dimension array *sizes*:

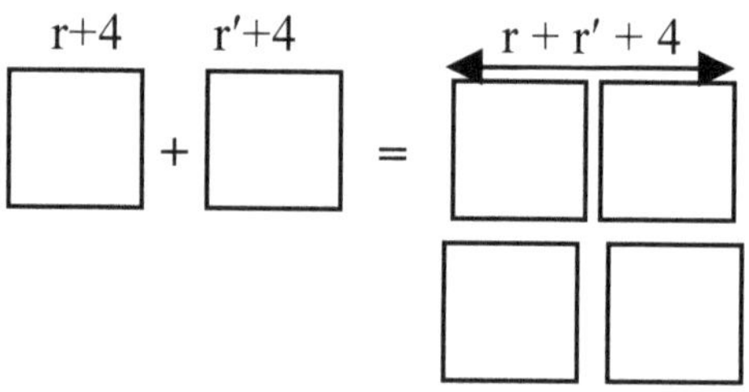

where the block sizes are multiplied by a factor of 4 or 3 as implied by eqs. 1.4.

1.1.2.2 Fractal Subtraction

Fractal subtraction causes removal of parts of a fractal. Again we view it in the context of the dimension arrays of Cosmos spaces.

$$D(2^{r+4}) \text{ “–” } D(2^{r'+4}) = D(2^{r-r'})$$
$$D(2^{r+3}) \text{ “–” } D(2^{r'+3}) = D(2^{r-r'}) \tag{1.5}$$

based on

$$2^{r+4}2^{-(r'+4)} = 2^{r-r'} \tag{1.6}$$

$$2^{r+3}2^{-(r'+3)} = 2^{r-r'}$$

It is clear that dimension subtraction subtracts the powers of 2. Subtraction may be viewed in terms of dimension arrays as

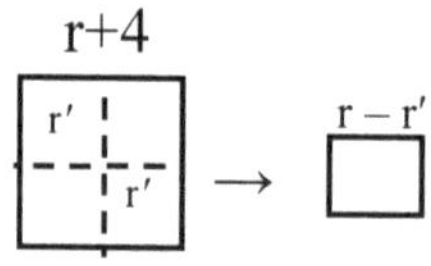

The interior r′ parts are excised by the subtraction. Fractal subtraction assumes r > r′.

1..1.2.3 Fractal Multiplication

We will define commutative fractal multiplication[5] of two fractals to yield the product of the fractal dimensions. An initial form of fractal multiplication can be implemented within the framework of Cosmos spaces:

---

[4] We have chosen to use base 2 in our study of dimensions because of its appearance in Cosmos spaces. Other choices of base are possible.

[5] We have replaced the non-commutative definition of fractal multiplication of *Fractal Cosmos* with this commutative definition.

$$D = D(2^{r+4}) \text{ “•” } D(2^{r'+4}) = (r + 4)(r' + 4) \tag{1.7}$$

For the multiplication of Cosmos dimension array fractals based on

$$2^{r+4} \text{ “•” } 2^{r'+4} = (2^{r+4})^{r'+4} = 2^{(r+4)(r'+4)} \tag{1.8}$$

with the result

$$D = \log_2(2^{(r+4)(r'+4)}) = (r' + 4)\log_2(2^{(r+4)}) = (r + 4)(r' + 4) \tag{1.9}$$

For 2nd Kind HyperCosmos dimension arrays

$$D = \log_2(2^{(r+3)(r'+3)}) = (r'+3)\log_2(2^{(r+3)}) = (r'+3)(r+3) \tag{1.10}$$

Fractal multiplication for Cosmos spaces takes every element of a dimension array and expands it (through multiplication) by the dimension array of the other multiplicand.

In general[6]

$$D = D(2^n) \text{ “•” } D(2^m) = mn \tag{1.7'}$$

1.1.2.4 Fractal Division

Fractal division[7] can be implemented within the framework of Cosmos spaces:

$$D = D(2^{r+4}) \text{ “/” } D(2^{r'+4}) \tag{1.11}$$

based on

$$2^{r+4} \text{ “/” } 2^{(r'+4)} = 2^{(r+4)/(r'+4)} \tag{1.12}$$

Then

$$D = \log_2(2^{(r+4)/(r'+4)}) = \log_2(2^{(r+4)})/(r' + 4) = (r+4)/(r' + 4) \tag{1.13}$$

In general[8]

$$D = D(2^m) \text{ “/” } D(2^n) = m/n \tag{1.13'}$$

Similarly for 2nd Kind HyperCosmos dimension arrays:

$$D = (r+3)/(r' + 3) \tag{1.14}$$

Fractal division of Cosmos spaces requires factoring within the r dimension array extracting r′ dimension array factors. Thus r > r′ is required or the result is a non-integer fractal dimension. Eqs. 1.13 and 1.14 explicitly show the factors.

---

[6] This expression may be generalized to powers greater than 2.

[7] We have replaced the definition of fractal division of *Fractal Cosmos* with this definition which is consistent with commutative fractal multiplication as defined above.

[8] This expression may be generalized to powers greater than 2.

1.1.2.5 Extension to Real Fractals

The addition of fractals in general combines the unit squares of two fractals in a manner similar to the illustration in section 1.1.2.1.

The subtraction of fractals in general separates the unit squares of the two fractals in a manner similar to the illustration in section 1.1.2.2. The terms in the fractal constructions must be separated to obtain the result. This presumes an extraction method (up to a multiple of the construction numbers) supporting their removal.

The multiplication of fractals in general combines the unit squares of two fractals in a manner similar to the description in section 1.1.2.3. The terms in one multiplicand of its fractal construction must be multiplied by the terms of the construction of the second multiplicand. The result is a combined fractal with terms being fractals of the second multiplicand.

The division of fractals in general decomposes the unit square of one fractal piecewise linear curve by piecewise linear curve extraction of the divisor fractal. The terms in the dividend in its fractal construction are divided by the terms of the construction of the divisor.

1.2 The Fractal Curve of the Cosmos – Double Threaded

The Cosmos Curve begins with a zero dimension point, and grows to a one dimension piecewise linear line segment by a simple aggregation of infinitesimal elements based on the identity

$$\sum_{k=\infty}^{1} 2^{-k} = 1 \tag{1.15}$$

It starts from a point where $k = \infty$, initially adding infinitesimal increments, which then sum to a unit line segment. The order of the summation begins with the infinitesimal and proceeds to the larger summands.

Then the unit piecewise linear line segment grows through a construction consisting of steps to form a fractal line. At the nth stage of the construction the growing line has the length 2^n. This line is compressed to fit within a unit square of sides 1 by 1. The line becomes fractal in the limit $n \rightarrow \infty$. The resulting fractal line is known as a Hilbert curve and, in the present context, the Cosmos Fractal Curve.

1.2.1 The Steps of the Construction

The steps of the construction process display a feature that relates to the distinction between HyperCosmos spaces and 2nd Kind HyperCosmos spaces.[9] The feature follows from our requirement that r be an even integer.[10]

For *even* n we have the piecewise linear segment of length 2^n that corresponds to a HyperCosmos space of dimension r whose dimension array size is 2^{r+4}. We can map

---

[9] A HyperCosmos space has both a Normal and a Dark sector. A 2nd Kind HyperCosmos space only has a Normal sector. As a result their dimension array sizes differ by a factor of 2.

[10] The value of r is even to prevent double counting of spaces since an r+1 dimension array is the same as an r dimension array.

the linear segment length to the dimension array size by linearizing the dimension array square into a sequence of dimensions. The result is

$$2^n = 2^{r+4} \qquad (1.16)$$

Then

$$n = r + 4 \qquad (1.17)$$

In terms of dimensions we see

$$D(2^n) = D(2^{r+4}) \qquad (1.18)$$

for a HyperCosmos dimension array map.

For *odd* n we have a piecewise linear segment of length 2^n. In order to preserve the oddness of n we must relate it to a 2nd Kind HyperCosmos space of dimension r whose dimension array size is 2^{r+3}. Again we can map the linear segment length to the dimension array size by linearizing the dimension array into a sequence of dimensions. The result is

$$2^n = 2^{r+3} \qquad (1.19)$$

with

$$n = r + 3 \qquad (1.20)$$

or

$$D(2^n) = D(2^{r+3}) \qquad (1.21)$$

for a 2nd Kind HyperCosmos space. The odd n is matched by the odd value of r+ 3 of 2nd Kind HyperCosmos dimension arrays since r is an even integer.

1.2.2 The Sequence of Construction Steps

Given the above comments, we can define the construction steps in terms of HyperCosmos spaces as:

$$D(2^n) = \delta_{n,even}D(2^{r+4}) + \delta_{n,odd}D(2^{r+3}) \qquad (1.22)$$

or

$$D(2^n) = \cos^2(n\pi/2)\, D(2^{r+4}) + \sin^2(n\pi/2)\, D(2^{r+3}) \qquad (1.23)$$

We now let $\pi/2 \rightarrow$ the angle variable θ and examine the change in $D((2^n, \theta)$ as the angle changes. This angle variable allow a continuous progression between $D(2^{r+4})$ and $D(2^{r+3})$.

$$D(2^n, \theta) = \cos^2(n\theta)D(2^{r+4}) + \sin^2(n\theta)D(2^{r+3}) \qquad (1.24)$$

Note:

$$\cos^2(n\theta) + \sin^2(n\theta) = 1 \qquad (1.25)$$

We see for all n:

If \θ = 0, then $D(2^n, \theta) = D(2^{r+4})$ (1.26)
If \θ = π/(2n), then $D(2^n, \theta) = D(2^{r+3})$ (1.27)
If \θ = π/2, then $D(2^n, \theta=\pi/2) = \cos^2(n\pi/2)D(2^{r+4}) + \sin^2(n\pi/2)D(2^{r+3})$
giving oscillation with increasing n (1.28)

Eq. 1.22 shows that the construction steps of the fractal curve alternate between HyperCosmos space and 2nd Kind HyperCosmos space equivalents. We may view the construction process as alternating construction steps of HyperCosmos space, and 2nd Kind HyperCosmos space, piecewise linear segments.

Fig. 1.6 displays a part of the Cosmos Fractal Curve showing the "oscillation" between Cosmos and 2nd Kind Cosmos steps in n.

1.2.3 Quantum Interpretation of Oscillation in θ

Eq. 1.24 appears rather like a quantum probability expression with

$$\text{Probabilities: } \cos^2(n\theta) \text{ and } \sin^2(n\theta) \tag{1.29}$$
$$\text{Amplitudes: } e^{i\varphi}\cos(n\theta) \text{ and } e^{i\varphi}\sin(n\theta) \tag{1.30}$$

One might interpret eq. 1.24 as the relative probability of having a HyperCosmos space or a 2nd Kind HyperCosmos space for a given value of space-time dimension r.

We may therefore raise the possibility of a Quantum interpretation of the Fractal Cosmos Curve. The apparent problem is the lack of a Quantum interpretation of the space of the fractal curve, if any, within which the curve may reside. This issue is partly mitigated by the two dimension space generated by the space-filling Cosmos Fractal Curve and by the Schrödinger-like form of the Master Equation. It disappears when we realize the discourse above portrays a time order of Fractal Curve construction followed by the existence of the Master Equation and Cosmos spaces. Yet in the absence of time these processes are *coexistent* and exist as simultaneity. All occurs in an *Eternal Now*.

Since the curve, the two dimension space, the Master Equation and the Cosmos spaces spectrum are all *structural* with no basis in Reality since they have no mass or energy, they can be considered to exist as infinite structure for all times within universes. Time only exists within universes.

Eq. 1.30 can be interpreted as a probability amplitude that exists at the point of creation of a universe of space-time dimension r. It implies that a universe may be created having a HyperCosmos space structure with a dimension array of size 2^{r+4} or a 2nd Kind HyperCosmos space with a dimension array of size 2^{r+3} or a superposition of these space types. When a universe is instantiated it has one of these three possibilities. The difference between a HyperCosmos space structure and a 2nd Kind HyperCosmos space structure is the presence of a Dark sector for the first, and the absence of a Dark sector for the second, due to the factor of 2 difference in the dimension array sizes.

1.2.4 Quantum Interpretation of a Dark Sector in Our Universe

It is not certain whether a Dark sector exists for our universe. Yet there is indirect evidence for a Dark sector based primarily on gravitation studies. There is no convincing direct evidence. This dichotomy raises the possibility that our universe is a superposition of these space types. Then the Dark sector does not "exist" but the quantum effects of the superposition generate indirect "evidence" for a Dark sector based on quantum based gravitation effects.

1.3 Double Helix Fractal Curve

We note that the curves of values generated by the variations in θ have the appearance of a helix as shown in Fig. 1.7 for several reasons.

Firstly, the variation from θ = 0 continuously to θ = π/2n takes $D(2^n, \theta)$ from $D(2^{r+4})$ to $D(2^{r+3})$. This change corresponds to a change from n to n + 1 as shown in Fig. 1.7 with solid curving "helix" lines (the "front" view of the helix).

Secondly, the variation from θ = π/2n continuously to θ = 0 takes $D(2^n, \theta)$ from $D(2^{r+3})$ to $D(2^{r+4})$. This change corresponds to a change from n to n + 1 as shown in Fig. 1.7 with broken curving "helix" lines (the "rear" view of the helix).

Thirdly, the absence of spatial dimensions for the Cosmos Curve causes what might have been a truly helix curve to an oscillation between values as shown in Fig. 1.7. The construction step n may be viewed as a dimension of a space associated with the piecewise linear curves.

The sequence of construction of the Cosmos Fractal Curve alternates between HyperCosmos and 2nd Kind HyperCosmos space dimensions. The even and odd n construction steps support a separation of the Cosmos Curve into two strands that are interconnected step by step.

As a result the Cosmos Fractal Curve is the equivalent of a Double Helix curve. It thus is an analogue of DNA – the structure of organic life. See Fig. 1.8.

One may view the correspondence between the (flattened) double helix form of the Cosmos Curve in two dimensions, and the three dimension double helix form of DNA, as reflecting a similar firmness (solidity) of structure. The ubiquity of DNA in organic life reflects its essential strength as a Physical structure. DNA and the fractal Cosmos Curve are both long strands with an inherent stability.[11]

Our results support our contention that Cosmos Theory embodies a type of Physical Organism. The ubiquitous presence of fractals leads us to view the Cosmos as striated.

1.4 Extension of the Dimension Array Concept to a 2nd Space and a Space-time

By viewing a dimension array size as a fractal dimension, the possibility of extending our view of the Cosmos Curve appears. For each construction step n we can view the dimension array size[12] (2^{r+4} or 2^{r+3}) as the fractal (topological) dimension of its *associated Cosmos space*.

For each n a new second space also appears. This space exactly corresponds to the space-time dimension r. There is thus a coordinate space associated with each step n and its associated Cosmos space. One can view the Cosmos fractal construction process as also *generating both a Cosmos space and a space-time at each step n.*

The creation of a universe of the type of a Cosmos space r utilizes its associated space-time of dimension r. Cosmos spaces have both dimension arrays *and* space-times

[11] This author was perhaps the first to investigate the electronic structure of DNA using electron spin resonance (ESR) techniques in 1964.

[12] At step n we have n = r + 4 for n even, and n = r + 3 for n odd.

that are used in the formation of universes. *A universe of space type r thus becomes a fractal as well and has the dimension r.*

1.5 Second Kind HyperCosmos Spaces

The appearance of 2nd Kind HyperCosmos space steps in the Cosmos fractal construction corresponds to their appearance in the unification of Cosmos spaces. Fig. 1.4 shows the set of unification spaces of Cosmos Theory. This figure and the following Fig. 1.5 were presented in previous books by the author.

Fig. 1.5 shows the dimension array of the 42 dimension Full HyperUnification space. It has three parts: an A part for HyperCosmos spaces unification, a B part for 2nd Part HyperCosmos spaces unification, and a C part for a remaining dimension array.

Dimension array A part is 5,592,400×5,592,400 dimensions
Dimension array B part is 2,796,200×2,796,200 dimensions
Dimension array C part is 8×8 dimensions

The 2nd Kind HyperCosmos spaces appear here (and in parts of previous books) that correspond directly with the structure of the Cosmos Fractal Curve presented above. Thus we have a consistency between the Cosmos Fractal Curve construction and the analysis of the unification structure of Cosmos Theory spaces.

This chapter has shown the detailed relation of Cosmos Curve construction steps to HyperCosmos and 2nd Kind HyperCosmos spaces. It has also associated a Cosmos space with (topological) dimension, and also with a space-time for each construction step n.

THE HYPERCOSMOS SPACES SPECTRUM

Blaha Space Number $N = o_s$	Cayley-Dickson Number n	Cayley Number Cn d_c	Dimension Array column length d_{cd}	Dimension Array Size $d_{dN}=d_{dr}$	Space-time-Dimension r	CASe Group $su(2^{r/2},2^{r/2})$ CASe
0	10	1024	2048	2048^2	18	su(512,512)
1	9	512	1024	1024^2	16	su(256,256)
2	8	256	512	512^2	14	su(128,128)
3	7	128	256	256^2	12	su(64,64)
4	6	64	128	128^2	10	su(32,32)
5	5	32	64	64^2	8	su(16,16)
6	4	16	32	32^2	6	su(8,8)
7	**3**	**8**	**16**	$\mathbf{16^2}$	**4**	**su(4,4)**
8	2	4	8	8^2	2	su(2,2)
9	1	2	4	4^2	0	su(1,1)

Figure 1.1. The HyperCosmos space spectrum.

THE Limos SECTOR SPACES SPECTRUM

Blaha Space Number $N = O_s$	Cayley-Dickson Number n	Cayley Number d_c	Row Size d_{cr}	Dimension Array Size $d_{dN}=d_{dr}$	Space-time-Dimension r	CASe Group $su(2^{r/2},2^{r/2})$ CASe
10	0	1	2	2^2	-2	U(1)
11	-1	½	1	1^2	-4	U(½)
12	-2	¼	½	$½^2$	-6	U(¼)
13	-3	1/8	¼	$¼^2$	-8	U(1/8)
14	-4	1/16	1/8	$1/8^2$	-10	U(1/16)
				•		
				•		
				•		
∞	$-\infty$	0	0	0	$-\infty$	U(0)

Figure 1.2. The Cosmos Theory space spectrum Limos spaces for Gambol Theory. See Blaha (2023e).

HYPERCOSMOS OF THE SECOND KIND SPACES SPECTRUM

Blaha Space Number $N = O_s$	Cayley-Dickson Number n	Cayley Number d_c	Dimension Array size $d_{dN2} = d_{dr2}$	Space-time-Dimension r	CASe Group $su(2^{r/2}, 2^{r/2})$ CASe
0	10	1024	1024 × 2048	18	su(512,512)
1	9	512	512 × 1024	16	su(256,256)
2	8	256	256 × 512	14	su(128,128)
3	7	128	128 × 256	12	su(64,64)
4	6	64	64 × 128	10	su(32,32)
5	5	32	32 × 64	8	su(16,16)
6	4	16	16 × 32	6	su(8,8)
7	**3**	**8**	**8 × 16**	**4**	**su(4,4)**
8	**2**	4	4 × 8	2	su(2,2)
9	**1**	2	2 × 4	0	su(1,1)

Figure 1.3. Part of the HyperCosmos of the Second Kind space spectrum.

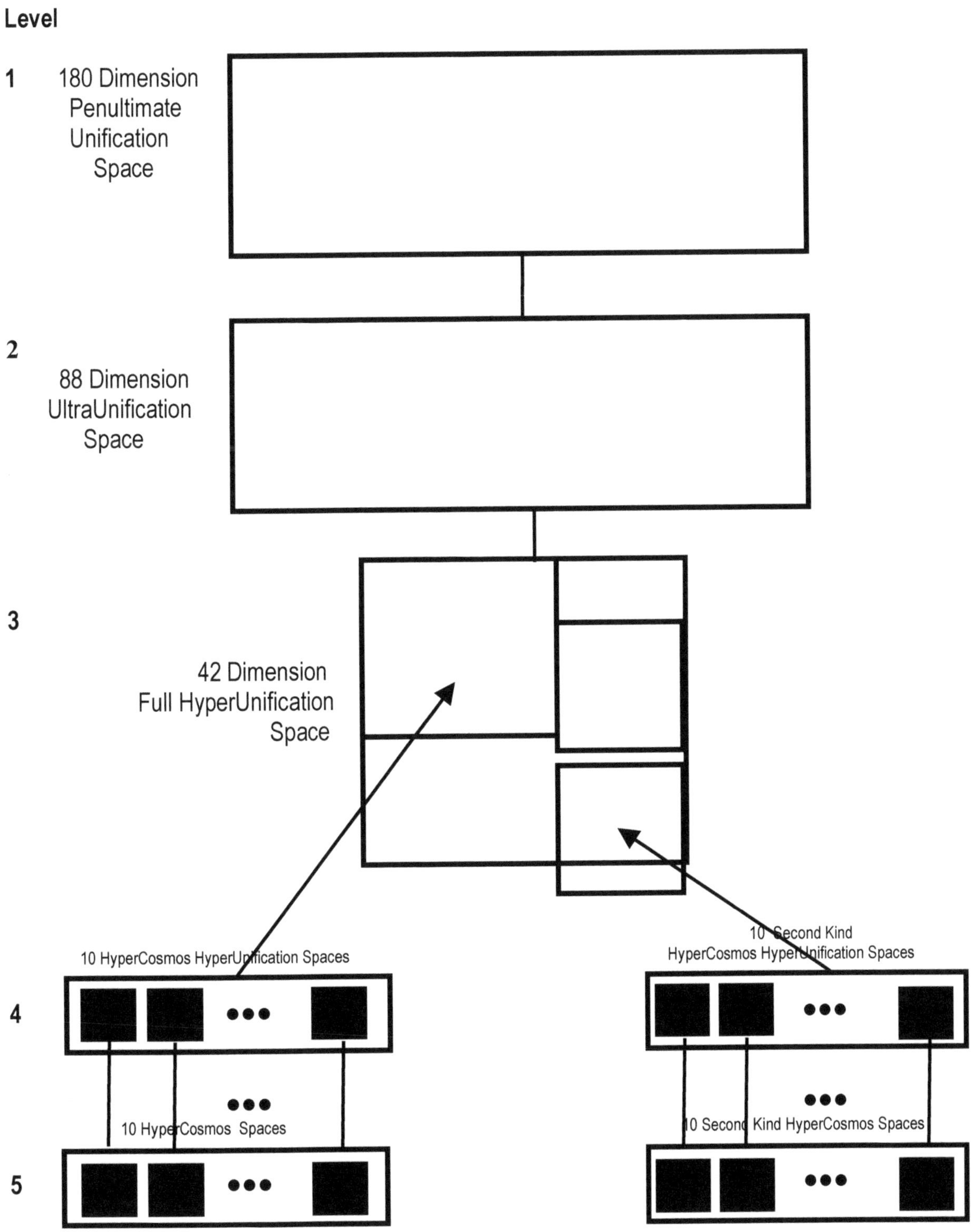

Figure 1.4. Diagram of the five levels of the Cosmos spaces unification. They contain 43 spaces.

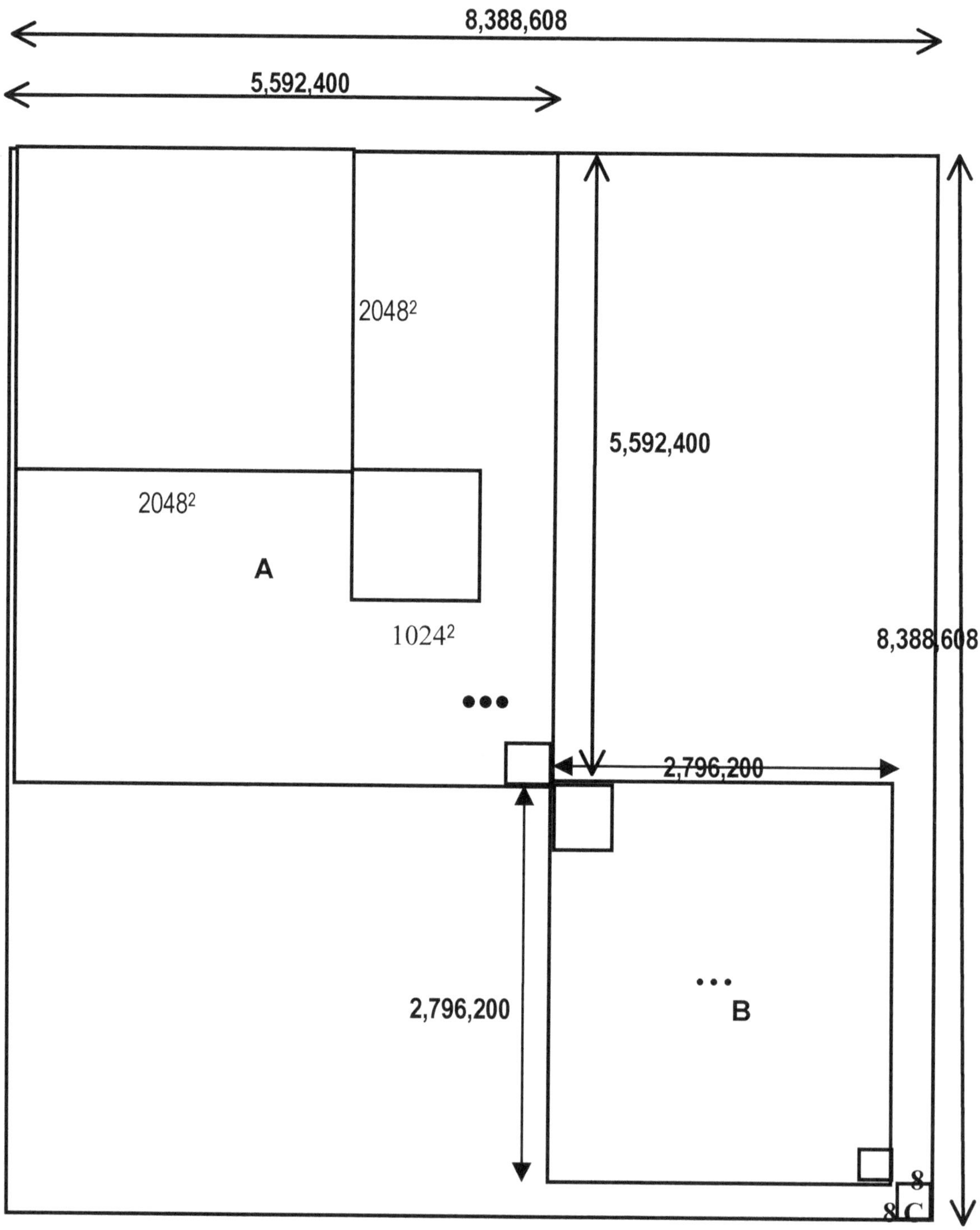

Figure 1.5. Form of the 42 space-time dimnesion of the Full HyperUnification space. All blocks are square. The figure is not drawn to scale.

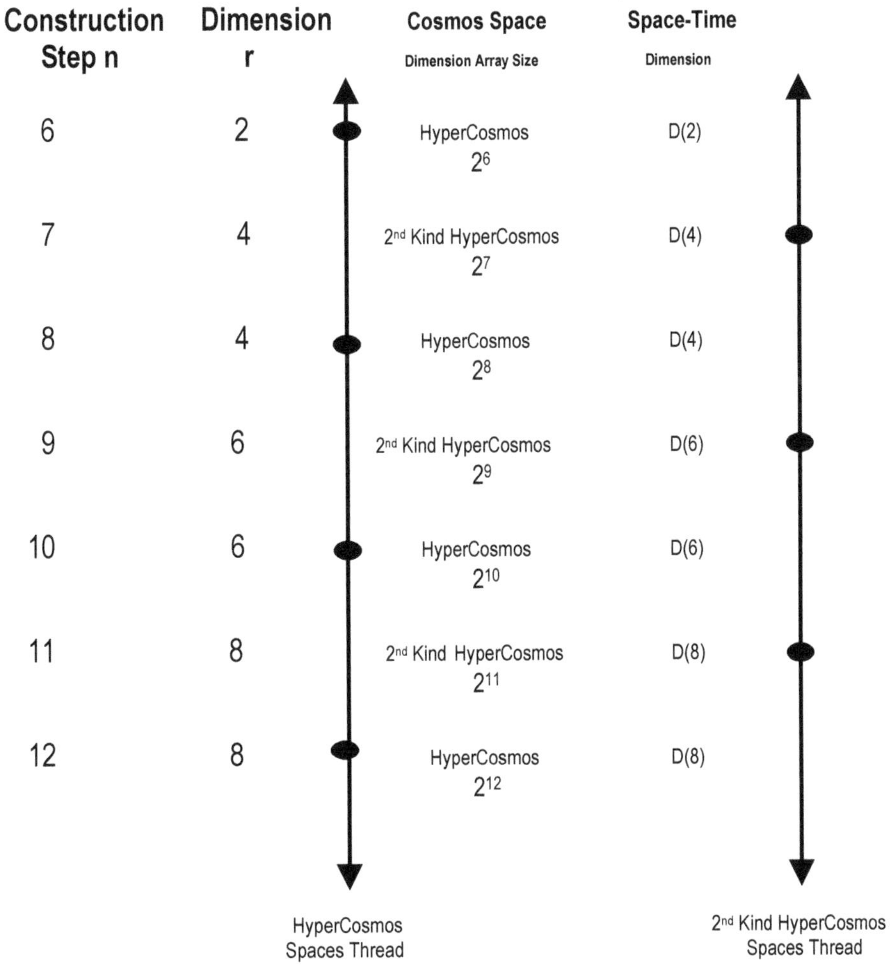

Figure 1.6. Section of the Cosmos Curve showing the two threads of Cosmos spaces and space-times. The space-times D(r) for each n is based on eq. 1.1. The piecewise linear line length is 2^n. At each construction step n there is the value of the linear line, its Cosmos space equivalent, and the space-time equivalent. Note the repetitions of space-time dimensions is correct since it reflects the presence of both HyperCosmos and 2nd HyperCosmos spaces in the two threads. The "dots" indicate the type and presence of each Cosmos space on each thread for that value of n. Note the alternation between threads as n increases. The powers of 2 are the dimension array sizes.

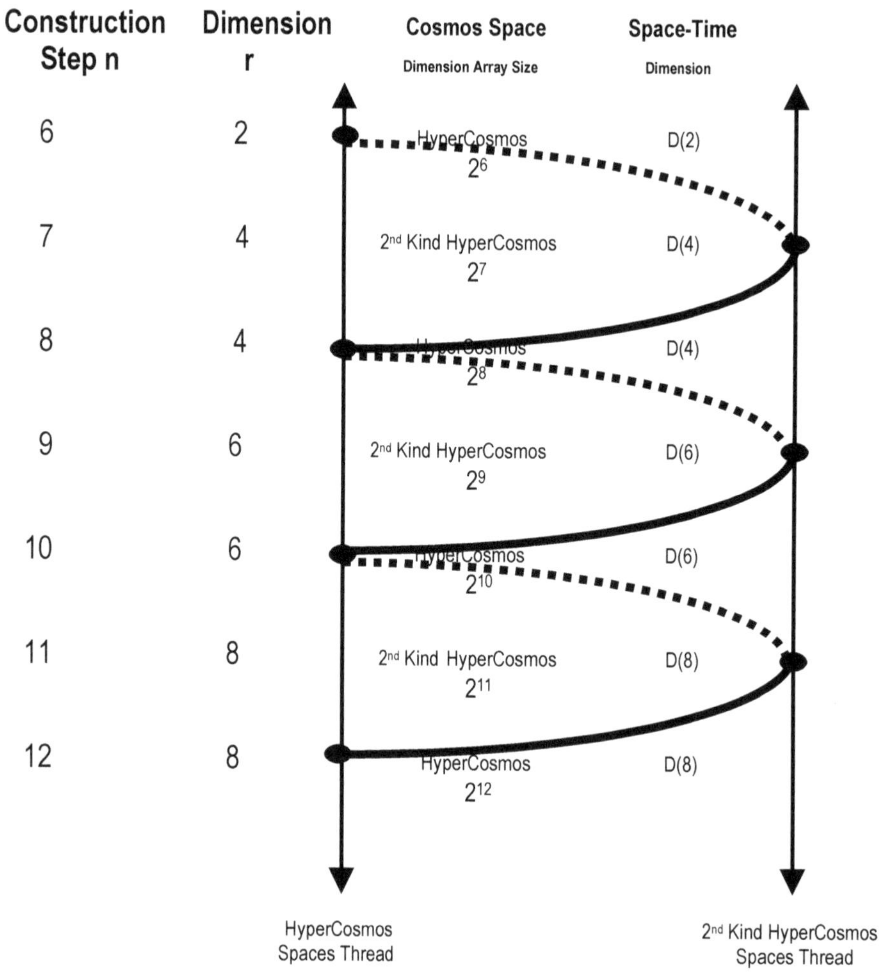

Figure 1.7. Section of the Cosmos Curve showing a double helix form (with the helix mapped to two dimensions.) The two threads of Cosmos spaces and space-times with the "helix" connection between strands are due to the variation in θ. The connecting curved lines form a helix when the 0 spatial dimension of the cutve is considered – an analogue of a double helix similar to DNA. The variation from $\theta = 0$ continuously to $\theta = \pi/(2n)$ takes $D(2^n, \theta)$ from $D(2^{r+4})$ to $D(2^{r+3})$. This change corresponds to a change from n to n + 1 as shown above with solid curving "helix" lines (the "front" view of the helix). The variation from $\theta = \pi/(2n)$ continuously to $\theta = 0$ takes $D(2^n, \theta)$ from $D(2^{r+3})$ to $D(2^{r+4})$. This change corresponds to a change from n to n + 1 as shown above with broken curving "helix" lines (the "rear" view of the helix). The curved lines are not drawn to scale.

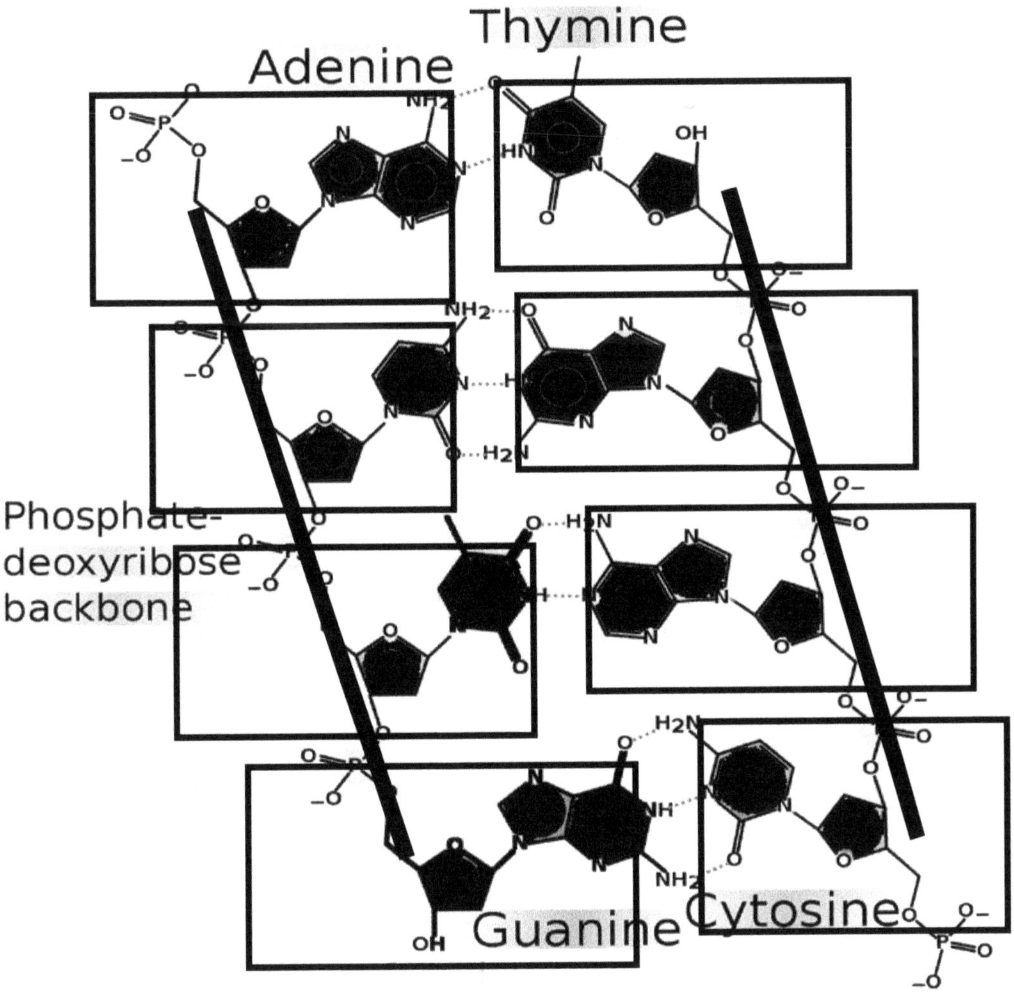

Figure 1.8. The Cosmos double helix structure superimposed on a four base pair fragment of double helix DNA. The correspondence is evident. From Fig. 5.3 of Blaha (2021e).

2. Fractal Representation of Cosmos Space Sequences

In chapter 1 we developed a fractal view of the construction steps in the definition of the Cosmos Fractal Curve. At each step n we associated a fractal with a dimension array of $2^n = 2^{r+4}$ or 2^{r+3} of a HyperCosmos space or a 2nd Kind HyperCosmos space respectively. As one increases the value of r by 2 (since r values are even integers) the dimension array size increases by a factor of 4. We can represent this change with the equation:

$$2^{r+4}2^2 = 2^{r+6} = 2^{(r+2)+4}$$

or

$$D(2^{r+4}) + 4 = D(2^{r+2+4}) \quad \text{HyperCosmos space} \tag{2.1}$$

$$D(2^{r+3}) + 4 = D(2^{r+2+3}) \quad 2^{nd}\text{ Kind HyperCosmos space} \tag{2.2}$$

We now assume the quantity 4 that appears in these equations is a fractal dimension:

$$D(2^4) = 4 \tag{2.3}$$

by eq 2.2. We represent eqs. 2.1 and 2.2 as fractal additions:

$$\begin{aligned} D(2^{r+4}) \text{ “+” } D(2^4) &= D(2^{r+2+4}) \\ D(2^{r+3}) \text{ “+” } D(2^4) &= D(2^{r+2+3}) \end{aligned} \tag{2.4}$$

as described in chapter 1 in eq. 1.1 and also in eq. 2.4. Diagrammatically we represent eq. 2.4 as

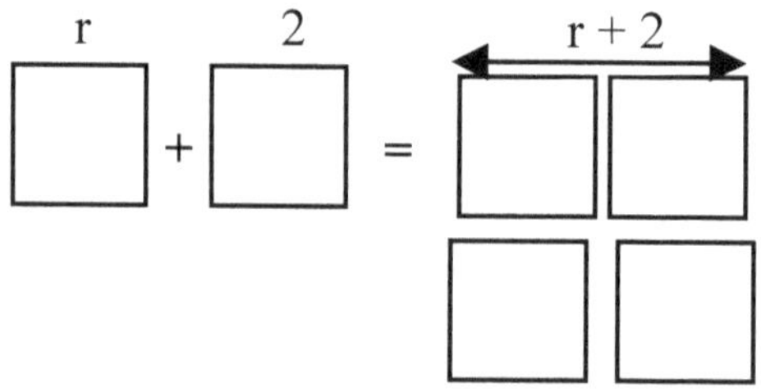

Figure 2.1. Addition of D(4) to a Cosmos Space Fractal. Note the D(4) signifies four items in a square array as shown with sides of 2. The D(4) indicates a dimension-type array of 4 entries and a row length of 2.

The r + 2 dimension array is generated. $D(2^4)$ represents 2^4 as a fractal using eq. 1.2:

$$D(2^4) = \log(2^4)/\log(2) = \log_2(16) = 4 \tag{2.5}$$

Eq. 2.4 shows the fractal $D(2^4)$ represents a multiplier of 4 for the dimension array for r extended to r + 2.

2.1 Dimension Array Column Length

If one extends the dimension array row (and column) length by 2 dimensions then one has the fractal addition:

$$D(r) \text{“+”} D(2^2) = D(r+2) \tag{2.6}$$

which corresponds to the multiplier “2” in the A1 row of Fig. 2.2 for column lengths:

A1 Cosmos spaces dimension array column lengths $2^{r/2+2}$	4	**2** = D(4) = $\mathbf{ln_2(4)}$	20.79

2.2 The Multiplier is a Fractal

The appearance of fractal addition in eqs. 2.4 suggests that the addend $D(2^4)$ is also is associated with a fractal dimension. When we consider the construction steps for the Cosmos Fractal Curve we find the fractal Cosmos space for dimension r increases to the fractal Cosmos space of dimension r + 2 for each construction step.

In particular, Fig. 2.6 shows that the steps for even values of n correspond to Cosmos spaces increase by 2, and also the steps for odd values of n correspond to 2nd Kind Cosmos spaces increase by 2.

The increase for Cosmos space steps (even n) may be viewed as

$$D(2^n) \text{“+”} D(2^2) = D(2^{r+4}) \text{“+”} D(2^2) = D(2^{r+6}) = D(2^{n+2}) \tag{2.7}$$

based on n = r + 4 and

$$2^n 2^2 = 2^{r+4}2^2 = 2^{(r+2)+4} = 2^{r+6} = 2^{n+2}$$

For odd n

$$D(2^n) \text{“+”} D(2^2) = D(2^{r+3}) \text{“+”} D(2^2) = D(2^{r+5}) = D(2^{n+2}) \tag{2.8}$$

based on n = r + 3 and

$$2^n 2^2 = 2^{r+3}2^2 = 2^{r+5} = 2^{(r+2)+3} = 2^{n+2}$$

Thus we can see that the increments $D(2^2)$, which correspond to multipliers of 4 for Cosmos dimensions r, have a fractal formulation. Thus we suggest all multipliers for the Cosmos Fractal Curve are fractal.

Descending the construction steps is possible through the use of fractal subtraction:

$$D(2^n) \text{“–”} D(2^2) = D(2^{r+4}) \text{“–”} D(2^2) = D(2^{r+2}) = D(2^{n-2}) \tag{2.9}$$
$$D(2^n) \text{“–”} D(2^2) = D(2^{r+3}) \text{“–”} D(2^2) = D(2^{r+1}) = D(2^{n-2}) \tag{2.10}$$

for the HyperCosmos (even n) and 2nd Kind HyperCosmos (odd n) cases respectively.

2.3 All Multipliers are Fractal

The fractal nature of the Cosmos Fractal Curve multipliers seen above leads us to state all multipliers in Fig. 2.2 are fractal. We indicate this fact in Fig. 2.2 by showing the fractal logarithm value for each multiplier.

2.4 Extension to Coupling Constants and Fermions

In later chapters we will show that coupling constant and fermion mass sequences, which use Fig. 2.2 multipliers are also necessarily fractal. Their fractal nature raises profound questions of the origin of their fractality.

Sequence Fractal	n_φ	g_g Fractal Dimension multiplier factor	λ_{n_φ}
Fractal	12	$1{,}26 = \ln_3(4)$	187.1
Fractal	10	$1.317 = \ln_{2.7}(3.7) = \ln_5(8)$	129.9
Fractal	8	$1.41 = \ln_{2.36}(3.36) = \ln_4(7)$	83.16
Sierpinski Triangles Fractal; 1.58-bit Large Language Model	6	$1.585 = \ln_2(3)$	46.778
A1 Cosmos spaces dimension array column lengths $2^{r/2+2}$ And SU(n) dimension array column lengths 2^n	4	$2 = \ln_2(4)$	20.79
A2 Cosmos unification space dimension r	4	$2 = \ln_2(4)$	20.79
A3 The sequence of Unification Spaces	4	$2 = \ln_2(4)$	20.79
B Coupling Constants α Multiplier	2	$4 = \ln_2(16)$	5.39
C Fermion Sequence 2 Masses	1	$2/3 \times 2^5 = \ln_2((32^{32})^{0.133})$	1.3
D Fermion Sequence 1 Masses	$½^{½}$	$3/2 \times \pi 2^5 = \ln_2((32^{32})^{0.3\pi})$	0.65
E n_φ Values Sequence; A Sequence of Sequences	½	$95939 = 5.96 \ln_2(32^{32\cdot 32\pi})$	0.325
F New n_φ Values Sequence based on α_{QED}	$0.48 \approx 2^{½}/3$	$844242 = 5.15 \ln_2(32^{32\cdot 32\cdot 32})$	0.303

Figure 2.2. Extended Cosmos Theory Sequences with a sequence F determined by setting λ = 0.303, which is the value of the QED couping constant e which determines the Fine Structure Constant α. The first row is for the case of a multiplier √2. Its n_φ value is an extension of the series of n_φ values 2, 4, 8 of rows A1 through B. The first and second rows are for the cases $n_\varphi = 6$ and $n_\varphi = 8$ which also may be of interest. The second row may correspond to 1.58-bit Large Language Models. Note: $32\cdot 32 = 32^2$ and $32\cdot 32\cdot 32 = 32^3$ above.

3. Fractal Representation of Coupling Constants

The sequence of coupling constants appears in Fig. 3.1. The coupling constants α_i have a clear multiplier – a factor of 4 – corresponding to row B of Fig. 2.2: Their multiplier is a fractal number – a fractal (topological in this case) dimension.

B Coupling Constants α Multiplier	2	$4 = \ln_2(16)$	5.39

Consistency requires the values of coupling constants α_i are also fractal numbers. The multiplicative[13] relation between the members of the sequence is:

$$D(2^{\alpha_i}) \text{ “}\bullet\text{” } D(2^4) = D(2^{\alpha_{i+1}}) \tag{3.1}$$

based on

$$2^{\alpha_i} \text{ “}\bullet\text{” } 2^4 = 2^{\alpha_{i+1}} \tag{3.2}$$

$$\log_2(2^{4\alpha_i}) = \alpha_{i+1} \tag{3.3}$$

yields eq. 3.1.

The coupling constants also satisfy fractal division:

$$D(2^{\alpha_i}) \text{ “/” } D(4) = D(2^{\alpha_{i-1}}) \tag{3.4}$$

based on

$$2^{\alpha_i} \text{ “/” } 4 = 2^{\alpha_i/2} \tag{3.4'}$$

We view the multipliers between α_i values as D(4) fractal dimensions in themselves. Then the actual α_i values are also fractal dimensions. *Thus there are six fractal coupling constants at present.*

[1313] See eqs. 2.9 and 2.10 for the general case of multiplication.

	ESTIMATE			EXPERIMENT		
Interaction	Expression	$\alpha_i = g^2/4\pi$ Value	g	Known[14] Value $g^2/4\pi$		Deviation
U(0) α_0	$e^2/4096$	0.0018036	-	-		-
U(1) $\alpha_1 = \alpha$	$e^2/1024$	0.00721438	0.303	0.0072973525643		1.15%
SU(2) $\alpha_2 = g^2/4\pi$	$e^2/256$	0.0289	0.63	0.0316		9.3%
SU(3) $\alpha_3 = \alpha_S$	$e^2/64$	0.115	1.21	0.117		1.7%
SU(4)[15] α_4	$e^2/16$	0.462	2.4?	0.458		0.087%

Figure 3.1. Coupling constants for the UST (and the Standard Model). Note factor of four relation between adjoining α values.[16]

3.1 The Coupling Constant Algorithm

The coupling constant counting algorithm is described in Blaha (2018e) and (2024i). Its form is consistent with the fractal representation with the multiplier 4 described above. The algorithm is:

> We have shown coupling constants appear to form a powers sequence. We now consider the expression of a coupling constant as the product of a number of factors that are based on geometry.
>
> First we note that n is the number of fermions in a fundamental representation of SU(n). The coupling constant $g(n) = g_n$ has the form:
>
> $$g(n) = e\,(4\pi)^{1/2}\,2^{n-6} = e\,(2^{-10}\pi)^{1/2}\,2^n = 0.1505\;2^n \qquad (1.18)$$
> $$= g_g\,(\text{number of spin states per fermion})^{\text{number of fermions}}$$
> $$= g_g\,(\text{number of spin states per fermion})^n$$
>
> n is the number of fermions in irreducible representation
>
> $$= g_g\,2^n$$
>
> The coupling constant is a fundamental value g_g times the number of spin states (a geometric property of fermions equal to half the column length of a spinor) raised to the number of fermions. The number of spin states per fermion is 2 in r = 4 dimensions.
>
> We see that g(n) has n factors – one factor for 2 spin states for each fermion in the SU(n) fundamental representation. The origin of g_g appears to be
>
> $$g_g = e\,(2^{-10}\pi)^{1/2} \qquad (1.19)$$
>
> Note
>
> $$\alpha_n = g_n^2/4\pi = e^2\,2^{2n-12}$$
> $$= \alpha_0\,2^{2n} \qquad (1.20)$$

---

[14] All coupling constant values are based on data from Particle Data Group Tables of 2024.

[15] This value is based on the "doubling trend" seen in the three known coupling constants g_i above in Fig. 3.2.

[16] From Blaha (2018e).

Eq. 1.18 is *initially* not decisively understood due to the appearance of the natural logarithm base e. However e and π naturally appear later in our studies of higher dimension volumes and surface areas. We therefore view g_g as having a *geometric origin.*

Eq. 1.18 implies that the coupling g(n) for SU(n) is g_g times n factors of 2 which represents the number of spin states per fermion in the SU(n) fundamental representation in four dimensions. Thus it is a product of degrees of freedom.

An important point is the number of factors is equal to the number of fermions in the SU(n) irreducible representation. *The coupling constant "felt" by one fermion depends on the total number of fermions in the group's fundamental representation.*

3.2 Fractal Representation of the Coupling Constant Algorithm

The above algorithm may be simply expressed in terms of fractals. We begin by defining a fractal for SU(n) equal to 2 raised to the number of real-valued dimensions of the irreducible representation of the group:

$$D(SU(n)) = 2n \tag{3.5}$$

Then eq. 1.20 above has the fractal representation

$$D(2^{\alpha_0}) \text{ "•" } D(SU(n)) = D(2^{\alpha_n}) \tag{3.6}$$

giving

$$\alpha_n = \alpha_0 2^{2n} \tag{3.6'}$$

The coupling constants form a sequence labeled by n. Each element in the sequence corresponds to an SU(n) with a fractal dimension 2n. We associate the fractal dimension 2^n, taken as the length of a square dimension array, with a $(2n)^2$ group's irreducible representation and also a dimension array with *size* $d_s = 2^{2n}$ elements.

$$D(SU(n)) = \ln_2(2^{2n}) \tag{3.7}$$

Thus there is a sequence of SU(n)-related spaces just as there is a Cosmos spaces sequence.

3.3 SU(n) Spaces Dimension Arrays Applied to Internal Symmetry Spaces

The SU(n) dimension arrays with *size* 2^{2n} form a sequence. They can be used to subdivide a Cosmos space dimension array into parts. The possible parts for dimension arrays for internal symmetries are:

$$\begin{aligned}
&SU(1) \text{ - parts have } 2^2 = 4 \text{ real dimensions} \\
&SU(2) \text{ - parts have } 2^4 = 16 \text{ real dimensions} \\
&SU(3) \text{ - parts have } 2^{3•2} = 64 \text{ real dimensions} \\
&SU(4) \text{ - parts have } 2^{4•2} = 256 \text{ real dimensions} \\
&SU(5) \text{ - parts have } 2^{5•2} = 1024 \text{ real dimensions} \\
&\qquad \cdots
\end{aligned} \tag{3.8}$$

3.4 SU(n) Spaces Dimension Arrays Applied to Fermions

When we map a dimension array to fundamental fermions each dimension corresponds to one fermion. The dimension arrays of the internal symmetries can be used to layout the pattern of fermions in a dimension array. We followed that procedure in previous work.

We now consider the layouts determined by eq. 3.8 for the case of r = 4: our space-time which is described by our Unified Symmetry Group (UST). We note that the patterns can be nested. Figs. 3.1 – 3.4 show the patterns of SU(n) dimension arrays embedded within the UST fundamental fermion array.[17] These patterns can be nested to generate a well-defined UST fermion array spectrum.

3.5 SU(n) Spaces Dimension Arrays Applied to Symmetries

The discussion here may be applied to the patterns of symmetry irreducible representation dimensions as we presented in previous discussions of the UST. See Fig. 3.5.

---

[17] All figures of this chapter are from Blaha (2023d) and earlier books.

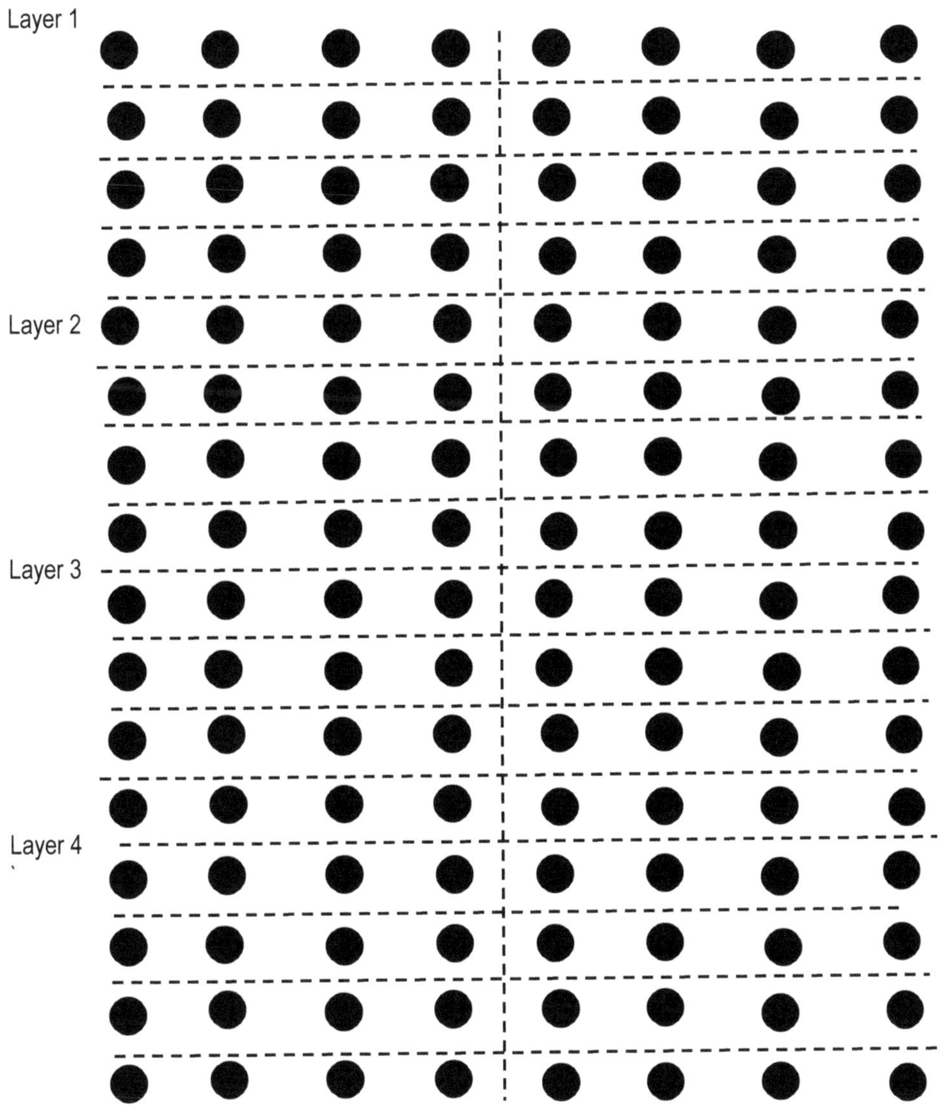

Figure 3.1. The four layers of the r = 4 dimension array applied to fundamental fermions. Only the Normal sector is shown. The Dark sector has a duplicate layout of Dark fermions. Note the fermions are partitioned into sets of four fermions as in the SU(1) line in eq. 3.8. This layout appears within the UST fermion dimension array.

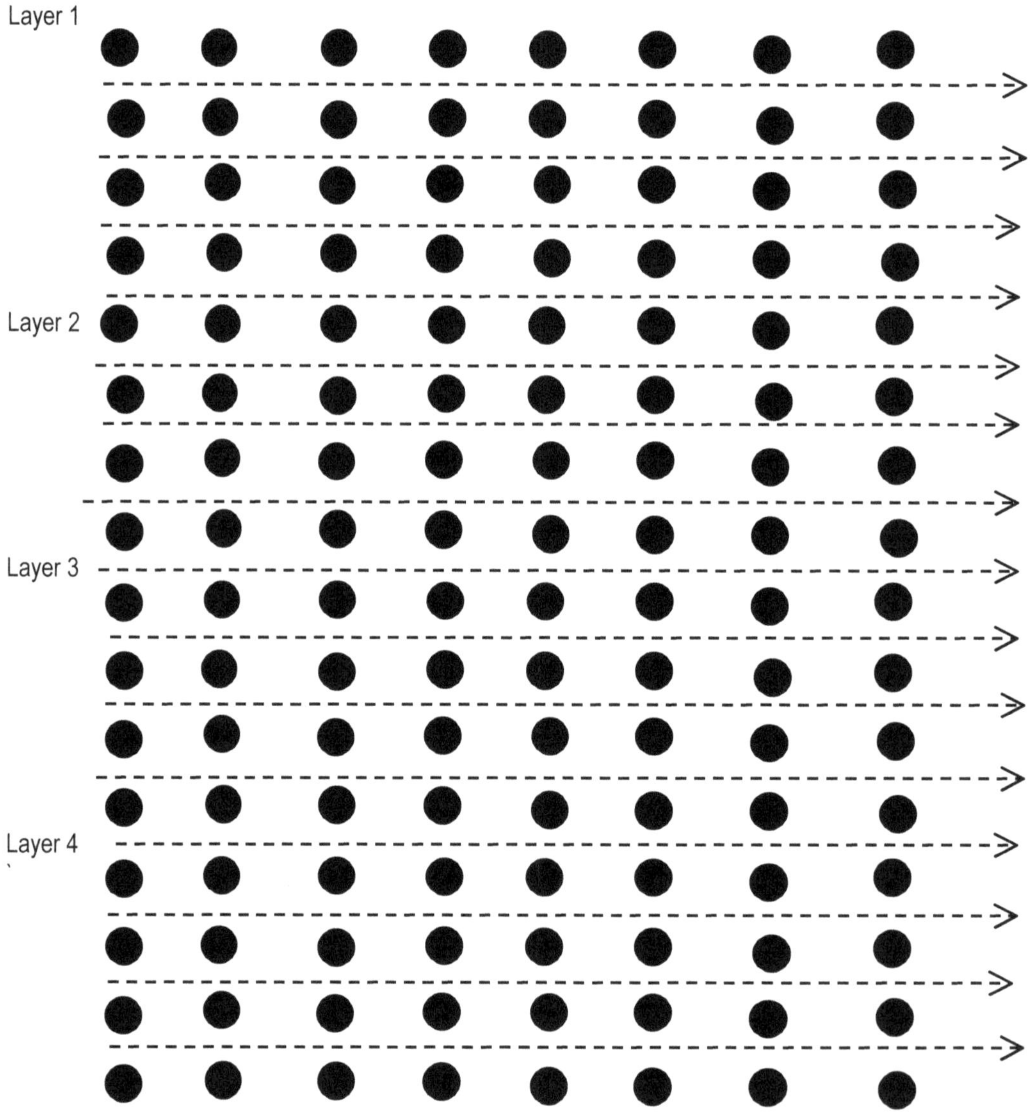

Figure 3.2. The four layers of the r = 4 dimension array applied to fundamental fermions. Only the Normal sector is shown. The Dark sector similarly extends the layout of Dark fermions. The fermions are partitioned into sets of sixteen fermions as in the SU(2) line in eq. 3.8 with the inclusion of the Dark sector parts indicated by the arrows. This layout appears within the UST fermion dimension array.

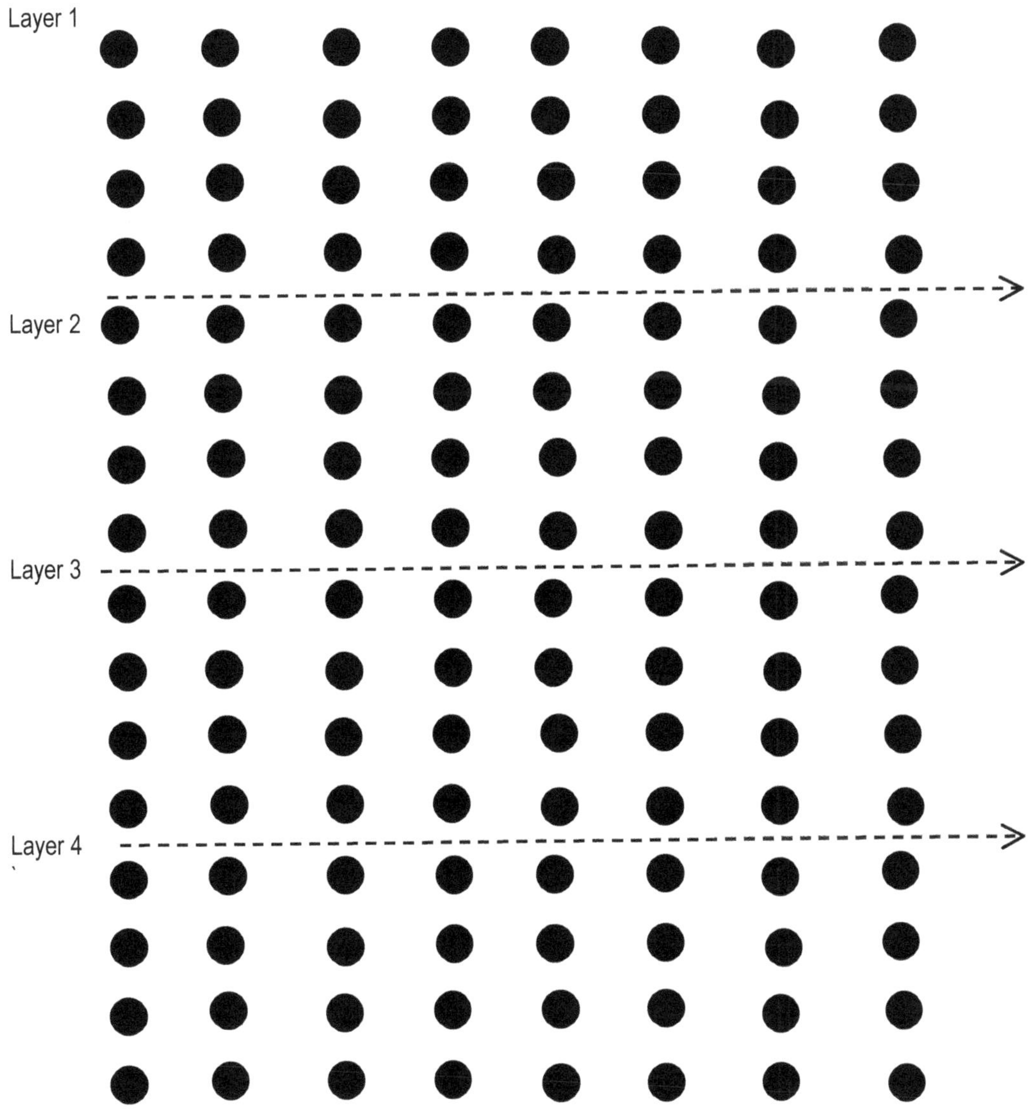

Figure 3.3. Four layers of the r = 4 dimension array with the Dark sector fermions not shown. Only the Normal sector is shown. The Dark sector similarly extends the layout of fermions. The fermions are partitioned into sets of sixty-four fermions as indicated in the SU(3) line in eq. 3.8 with the inclusion of the Dark sector parts indicated by the arrows. Each layer of Normal and Dark fermions has 64 fermions.This layout appears within the UST fermion dimension array.

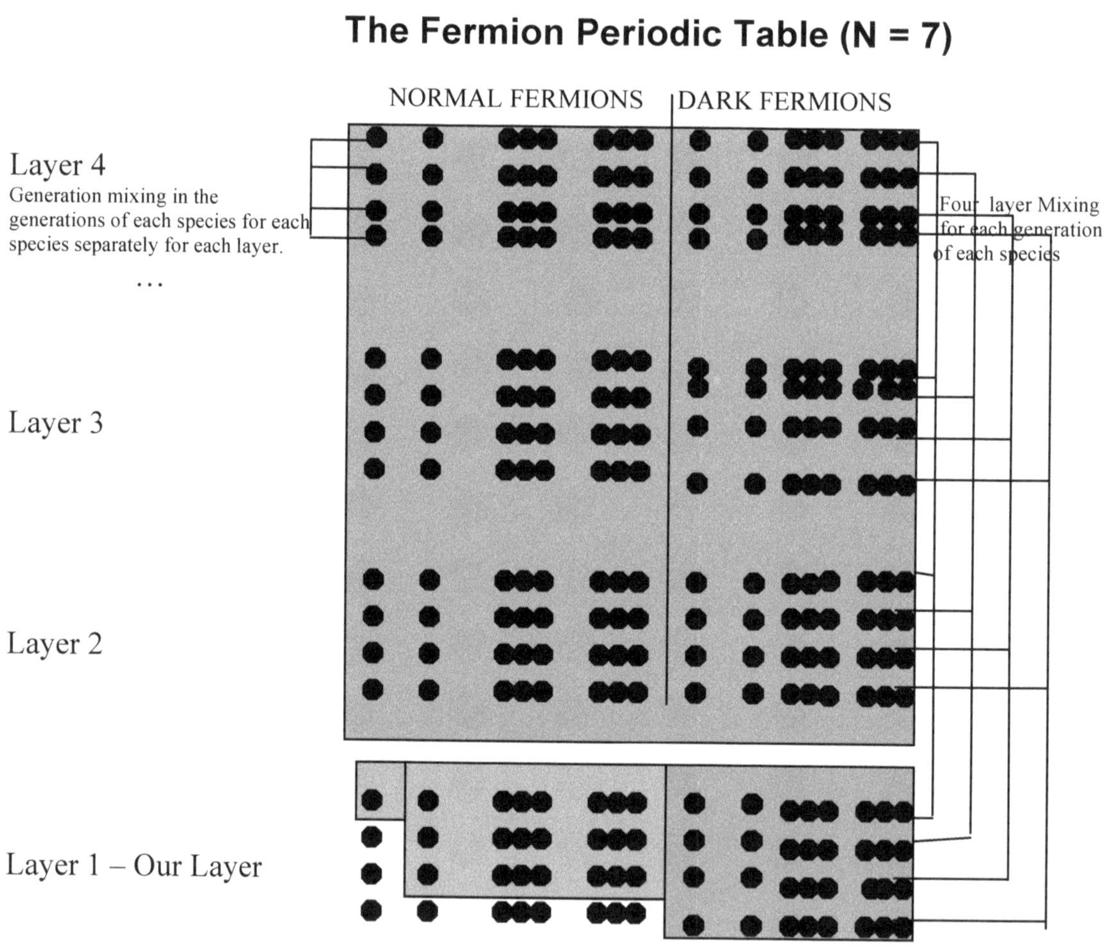

Figure 3.4. The UST fermions showing the nested partitioning implied by eq. 3.8 based on the *SU(n) dimension arrays*. The N = 7 QUeST Fermion particle spectrum and partial examples of the pattern of mass mixing of the Generation groups and of the Layer groups. Unshaded parts are the known fermions, A shaded generation is an additional, as yet not found, 4th generation. The lines on the left side (only shown for one layer) display the Generation Group mixing within each layer. The Generation mixing occurs within each layer using a separate Generation group for each layer. The lines on the right side show Layer group mixing (for Dark matter) with the mixing among all four layers for each of the four generations individually. There are four Layer groups for Normal matter and four Layer groups for Dark matter. There are 256 fundamental fermions.

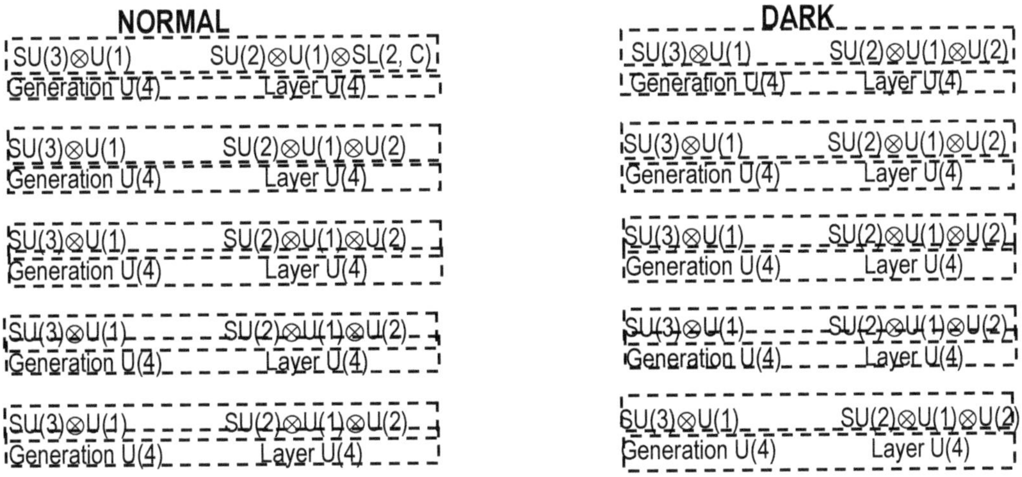

Figure 3.5. The sets of symmetries in the UST). Note each element has a 16 real dimension representation as in the SU(2) dimension array described in eq. 3.8. This depiction appears in the UST. The SL(2, **C**) representation has four dimensions.[18]

[18] The Lorentz Group $SO^{+}(1, 3)$ is often specified with an SL(2, **C**) representation.

4. Fractal Representation of the Coupling Constants Related to Fundamental Fermion Masses

This chapter relates the fractal representation Coupling Constants to Fundamental Representation Masses. In doing this, we will see that the fractal nature of Coupling Constants carries over to the multipliers and λ values for fermion mass sequences. In particular the λ values and the multipliers for the mass sequences (rows C and D of Fig. 2.2) are directly related to Coupling Constants.[19]

We conclude the fractal nature of Coupling Constants and these relations to mass sequences implies that the multipliers and λ values for fermion mass sequences are also *fractal*. Therefore fermion masses are fractal.

4.1 Cosmos Coupling Constant Values Related to Fundamental Fermion Masses

This section relates the coupling constant sequence to the fermion mass sequences.

4.1.1 The Fermion Masses Sequences

In Blaha (2024g), we found that the eight known fundamental fermions of the first layer of the four dimension Cosmos Theory space and UST could be separated into two sequences of four fermions. Sequence 2 which contained the down-type fermions was approximately estimated to have a series of ratios of $2^5 = 32$ factors between the masses. Sequence 1 fermions (up-type) had less consistent factors between masses. We tentatively estimated the ratio $2^5\pi = 32\pi$ based on an approximation to the t/c fermion mass ratio and to achieve a rough parallel with the down-type fermion factor.

The actual mass values of sequence 1 deviated significantly from the $2^5\pi$ multiplier. We attribute the deviations to additional perturbative interaction effects on masses that are beyond this model.

4.1.2 The Fundamental Fermion Masses Model

The mass sequences, as pictured below in Fig. 8.1, have a geometric character similar to the energy levels and the Cosmos spaces spectrum.

We now define a 2×2 Master Equation scaling equation for *both* sequences:

$$\begin{bmatrix} \partial^2/\partial u^2 + \lambda_1/u^2 - M_1 & 0 \\ 0 & \partial^2/\partial u^2 + \lambda_2/u^2 - M_2 \end{bmatrix} \begin{bmatrix} \varphi_1 \\ \varphi_2 \end{bmatrix} = 0 \qquad (4.1)$$

which leads to

[19] See Blaha (2025d) for more detail.

$$2^5\pi = \exp(\pi/(\lambda_1 - ¼)^{½}) \quad (4.2)$$
$$2^5 = \exp(\pi/(\lambda_2 - ¼)^{½}) \quad (4.3)$$

multipliers for the up-type and down-type sequences respectively seen in Fig. 2.2. Then

$$[(\ln(32\pi)/\pi]^{-2} + ¼ = = \lambda_1 = 0.714 \qquad \approx g_2 = 0.63 \quad (4.4)$$

$$[(\ln(32)/\pi]^{-2} + ¼ = = \lambda_2 = 1.07 \qquad \approx g_3 = 1.21 \quad (4.5)$$

where we adjoin Coupling Constant g values of comparable value from the coupling constant table (Fig. 3.1).

Note the values of λ_1 and λ_2 are related by:

$$1.07/.714 = 1.5 \quad (4.6)$$

Note the sum of the values $\lambda_1 + \lambda_2 = 1.784$ and also the sum $g_2 + g_3 = 1.84$. These values differ by 3.4% suggesting that the ElectroWeak SU(2) and Strong SU(3) interactions combine to generate each mass sequence through an interplay of interactions.

4.1.3 Separation of SU(2) and SU(3) Contributions to the Sequences' Powers

We view the values in the sequences as sums of SU(2) and SU(3) terms. We calculate the discrepancy x between the λ_i from

$$\lambda_1 + x = 1.21 \quad (4.7)$$
$$\lambda_2 - x = 0.63$$

using the approximate equality: $\lambda_1 + \lambda_2 = 1.784$ while $g_2 + g_3 = 1.84$. We find

$$x = 0.468 \quad (4.8)$$

.

After noting that $g_4 = 2.4$ is itself an extrapolation we will use the approximate values from Fig. 5.2 to obtain

$$x = ½(g_1 + g_2) = 0.466 \quad (4.9)$$

Then we find an approximate relation between the SU(1), SU(2) and SU(3) coupling constants, and the λ_i, factors in in the fundamental fermion mass sequences.

$$\lambda_1 = 0.714 = 1.21 - 0.468 = g_3 - ½(g_1 + g_2) \quad (4.10)$$

for sequence 1, and

$$\lambda_2 = 1.07 = 0.63 + 0.468 = 3g_2/2 + ½ g_1 \quad (4.11)$$

for sequence 2.

Sequence 2 is governed by the ElectroWeak interactions. Sequence 1 is governed by a combination of ElectroWeak and Strong SU(3) interactions. This chapter directly relates fermion mass sequences to the coupling constant sequence.

As a result we find the fermion masses and their sequence λ values are also fractal due to their relation to Coupling Constant fractals.

4.2 Strata Fermion Masses are Fractal Coupling Constant Dependent

Each fundamental fermion is one of four fermion analogues that differ only in mass. The analogues have identical features except for their masses and their stratum number. The masses differ by a factor of 95939 according to the entry in Fig. 3.2. The masses of each set of four corresponding fermions must be fractal based on the above arguments for fundamental fermion mass fractals due to their relation to Coupling Constant fractals and the 95939 multiplier fractal in .Fig. 3.2.

4.3 Gambol Fermion Masses are Fractal

Each fundamental fermion has a corresponding gambol fermion with the same quantum numbers except for their stratum number and gambol number. Their masses are a factor of 1/844242 of their corresponding fermion. This mass relation implies that they also are the result of the same process that generates the fundamental fermions' mass seen above. Therefore we apply the same logic as in section 4.1 to relate the gambol masses to the SU(3), SU(2) and U(1) coupling constants – thus making them fractal as well.

We conclude all fermion masses are fractal their λ values are fractal, and the 95939 and 844242 values are fractal.

5. Fractal Representation of Fermion Masses

We have seen that fermion masses scaled by some mass scale have fractal dimensions. We now consider the approximate mass sequences of Fig. 5.1 treated as fractal and derive their fractal interrelations.

The approximate masses for the two fundamental fermion sequences appear in Fig. 5.1. The four strata masses of each fundamental fermion type differ by a factor of 95939. The gambol fermion masses are lower by a factor of 844242 than their corresponding fermion's mass.

	SEQUENCE 1				SEQUENCE 2			
	e	u	c	t	ν'	d	s	b
Mass:	0.511×10^{-3}	1.80×10^{-3}	1.28	171	8.4×10^{-5}	4.24×10^{-3}	102×10^{-3}	4.34
Multiplier:	$2^5\pi$	$2^5\pi$	$2^5\pi$		32	32	32	

Figure 5.1 The fundamental fermion mass sequences with approximate multiplier factors between the masses of each sequence. The sequence 1 masses show a significant variation with their $2^5\pi$ factor but their overall trend is approximately satisfactory. Sequence 2 masses better approximate their factor of 32 progression.

The relations between the masses in Sequence 1 is multiplicative. When we consider their fractal dimensions, we see the logarithmic nature of dimensions causes the dimensions to undergo fractal multiplication in analogy to the Cosmos Fractal Curve construction steps in chapter 5. The fractal multiplier of $2^5\pi$ for sequence 1 takes a fermion mass to the next mass. We introduce a mass scale m_0 since fractal dimensions are dimensionless. We define the sequence 1 fractal mass ratios m_i

m_i: $m_e/m_0,\ m_u/m_0,\ m_c/m_0,\ m_t/m_0$

denoting them with an index i. Then

$$D(2^{m_i})\ \text{“•”}\ D(2^{32\pi}) = D(2^{m_{i+1}}) \qquad (5.1)$$

or

$$m_i \bullet 32\pi = m_{i+1} \qquad (5.1')$$

based on fractal multiplication as in eq. 1.7′

$$D = D(2^n)\ \text{“•”}\ D(2^m) = mn \qquad (1.7')$$

Similarly we find sequence 2 fractal mass ratios m_i

m_i: $m_{v'}/m_0$, m_d/m_0, m_s/m_0, m_b/m_0

also denoted with an index i gives

$$D(2^{m_i}) \text{ "•" } D(2^{32}) = D(2^{m_{i+1}}) \quad (5.2)$$

or

$$m_i \cdot 32 = m_{i+1}$$

These sequences are fractal in nature. *Each mass is fractal.* The multipliers are fractal.

5.1 Stratum Fractals

The strata fermion masses are separated by a factor of 95939 as shown in Fig. 5.2. Their fractal dimensions satisfy fractal addition as one ascends the strata

$$D(2^{m_i}) \text{ "•" } D(2^{95939}) = D(m_{i+1}) \quad (5.2)$$

or

$$m_i \cdot 95939 = m_{i+1} \quad (5.2')$$

where the index i labels the strata in the order of increasing masses.

5.2 Gambol Fractals

The gambol fermion masses are separated from their corresponding fermion by a factor of 844242 as shown in Fig. 5.2. Their fractal dimensions satisfy fractal division:

$$D(2^{m_{fermion}}) \text{ "/" } D(2^{844242}) = D(2^{m_{gambol}}) \quad (5.3)$$

or

$$m_{fermion}/844242 = m_{gambol} \quad (5.3')$$

5.3 The Nature of Fractal Fermions

We have represented fermion masses expressed in powers of 2 as fractals. In our previous books we have created a Gambol Theory in which each fundamental fermion in each stratum contains a set of gambols. We have chosen to identify the gambols with sets of fermion gambols defined as having 1/844242 of the fermion's mass with identical quantum numbers. Thus a fundamental fermion, and its various stratum siblings, are composed of 844242 gambol fermions. The gambol fermions within a fermion have a momentum distribution that we have suggested in previous books have a Planckian form.

On this basis we have suggested that gambol fermions are fractal as well.

5.4 What are the Properties of Fermions expressed in Terms of Gambol Fermions

We have developed a gambol theory of fundamental fermions that appears to match the known data such as deep inelastic lepton-nucleon scattering. We used Planckian distributions to model the internal gambol distributions.

We now consider a modified Planckian distribution in terms of fractal curve construction sequences. We will see the modified distribution has the scaling property characteristic of fractal curves. There is some support for believing the scaling property is present in fundamental fermion internals. For example, the electromagnetic vacuum polarization Γ(p) has been shown to have scaling in extensive calculations to all orders in perturbation theory:[20]

$$\Gamma_\mu(p) = f(\gamma_\mu + 2g\gamma\cdot pp_\mu/p^2)(p/\Lambda)^{2g} \tag{5.4}$$

$$\Gamma(p) = (p/\Lambda)^{2g} \tag{5.5}$$

where g is negative, f is a constant and Λ is a cutoff. Eq. 5.5 is manifestly scaling in the energy (momentum) p.

We now turn to a consideration of Planckian distributions seeking to identify a scaling form. We begin by recapitulating Einstein's derivation.

5.5 Einstein's Derivation of Planck Distribution for a Fixed Black Body

Einstein derived the Planck distribution following a statistical approach. Einstein's[21] derivation starts with the concept of quantum jumps between energy states $E_1 < E_2$ due to a radiation density u.[22]

There are three ways a quantum jump occurs: 1. spontaneous jumps with probability S_{12} for a transition from E_2 to E_1 with a quantum emission; 2. absorption of a quantum giving a jump from E_1 to E_2 with probability $A_{21}u$; and 3. induced emission of a quantum giving a jump from E_1 to E_2 with probability $I_{12}u$. The probabilities that states 1 and 2 are occupied in thermal equilibrium are proportional to $\exp(-E_1/kT)$ and $\exp(-E_2/kT)$ respectively. Thus

$$(S_{12} + I_{12}u)\exp(-E_2/kT) = A_{21}u\exp(-E_1/kT) \tag{5.6}$$

expresses the balance of transitions. The constants S_{12}, I_{12} and A_{21} are properties of the states and independent of the density u. At high temperature $S_{12} << I_{12}u$. This implies

$$I_{12} = A_{21} \tag{5.7}$$

The transitions: 1 → 2 and 2 → 1 induced by the electromagnetic field are the same.

The density u(E) then is

$$\begin{aligned} u(E) &= (S_{12}/I_{12})/(\exp((E_2 - E_1)/kT) - 1) \\ &= (S_{12}/I_{12})/(\exp((E/kT) - 1) \end{aligned} \tag{5.8}$$

---

[20] Stephen Blaha, Phys. Rev. **D9**, 2246 (1974). See Blaha (2025b) for a detailed account of Γ and the calculation of the Fine Structure Constant α to 13 places to all orders in perturbation theory. Other calculations of α typically calculate it based on questionable extrapolations of 4th order perturbation theory.

[21] A. Einstein, Verhandl, deut. Physic. Ges. **18**, 318 (1916); Physik. Z. **18**, 121 (1917); J. W. M. Dumond and E. R. Cohen, Revs. Mod. Physics **25**, 691 (1953).

[22] This section appeared in Blaha (2024a) and in his earlier books

Einstein's derivation assumes that the container and the radiation can coexist in a stable state and that transitions between states can be determined statistically.

5.6 A Gambol-based Derivation for Fundamental Fermions

We now consider a derivation of the energy distribution of gambols within a fundamental fermion. We note that Einstein's results, while good for macroscopic black bodies, appears to need modification when applied to the interior of fundamental fermions. In particular, Einstein's $u(E) \to 0$ as $E \to 0$ in eq. 5.8.

When we examine $\Gamma(p) = (p/\Lambda)^{2g}$ for *negative* g we find it becomes infinite at p = 0. This view is consistent with the finiteness of the John-Baker-Willey form of QED.

Therefore we propose to modify the states in the above derivation for occupation in thermal equilibrium to:

$$\exp(-E_1/kT) \to 1/(\exp(E_1/kT) - 1) \tag{5.9}$$

and

$$\exp(-E_2/kT) \to 1/(\exp(E_2/kT) - 1) \tag{5.10}$$

The consequences of this choice are:

1. Both occupation factors become infinite at $E_i = 0$ corresponding to $\Gamma(p)$ in eq. 5.5. The growth of occupation factors near E = 0 is consistent with the expectation of an infinity at ultra-short distances within a fermion.

2. Since $E_2 \geq E_1$ eqs. 5.10 and 5.9 are consistent.

In eq. 5.6 we then must replace

$$\begin{aligned} 1/\exp((E_2 - E_1)/kT) - 1) \to \quad & (\exp(E_1/kT) - 1)/\{(\exp(E_2/kT) - 1)\} \\ & = (1 - \exp(-E_1/kT))/(\exp((E_2 - E_1)/kT) - \exp(-E_1/kT) \end{aligned} \tag{5.11}$$

For large E_1 (and thus $E_2 \geq E_1$), we can approximate u(E) and obtain

$$\begin{aligned} &= 1/\exp((E_2 - E_1)/kT) = \exp(-(E_2 - E_1)/kT) \\ &= \exp(-E/kT) \end{aligned} \tag{5.12}$$

Comparing to eq. 5.6 we obtain

$$\begin{aligned} u(E) &= (S_{12}/I_{12})/(\exp((E_2 - E_1)/kT) - 1) \\ &= (S_{12}/I_{12})\exp(-E/kT) \\ &= (S_{12}/I_{12})(e^{E})^{-1/kT} \end{aligned} \tag{5.13}$$

Now we find $u(E) \to \infty$ as $E \to 0$ in a consistent manner.

The form of u(E) in eq. 5.13 has scaling in E in a manner expected of fractal curves. Note that the length of u(E) – measured along the curve – is infinite due to its infinite behavior as E → 0 – another characteristic feature of fractal curves. As one scales u(E) downward one sees a repetitive scale behavior – again a fractal feature.

We conclude the dynamic interiors of fundamental fermions are fractal. Since a fundamental fermion contains 844242 gambol fermions, a statistical representation of their internal dynamics is warranted.

Thus we have found the fractal form of the fermion internal energy distribution. It can now be used to do gambol calculations similar to

Cosmos Theory: The Sub-Particle Gambol Model (Pingree Hill Publishing, Auburn, NH, 2023).
and
Cosmos-Universe-Particle-Gambol Theory (Pingree Hill Publishing, Auburn, NH, 2024).

The preceding discussions lead us to propose that, not just masses, but each entire fermion particle is fractal and has a fractal substructure reflecting a deep inner dynamics.

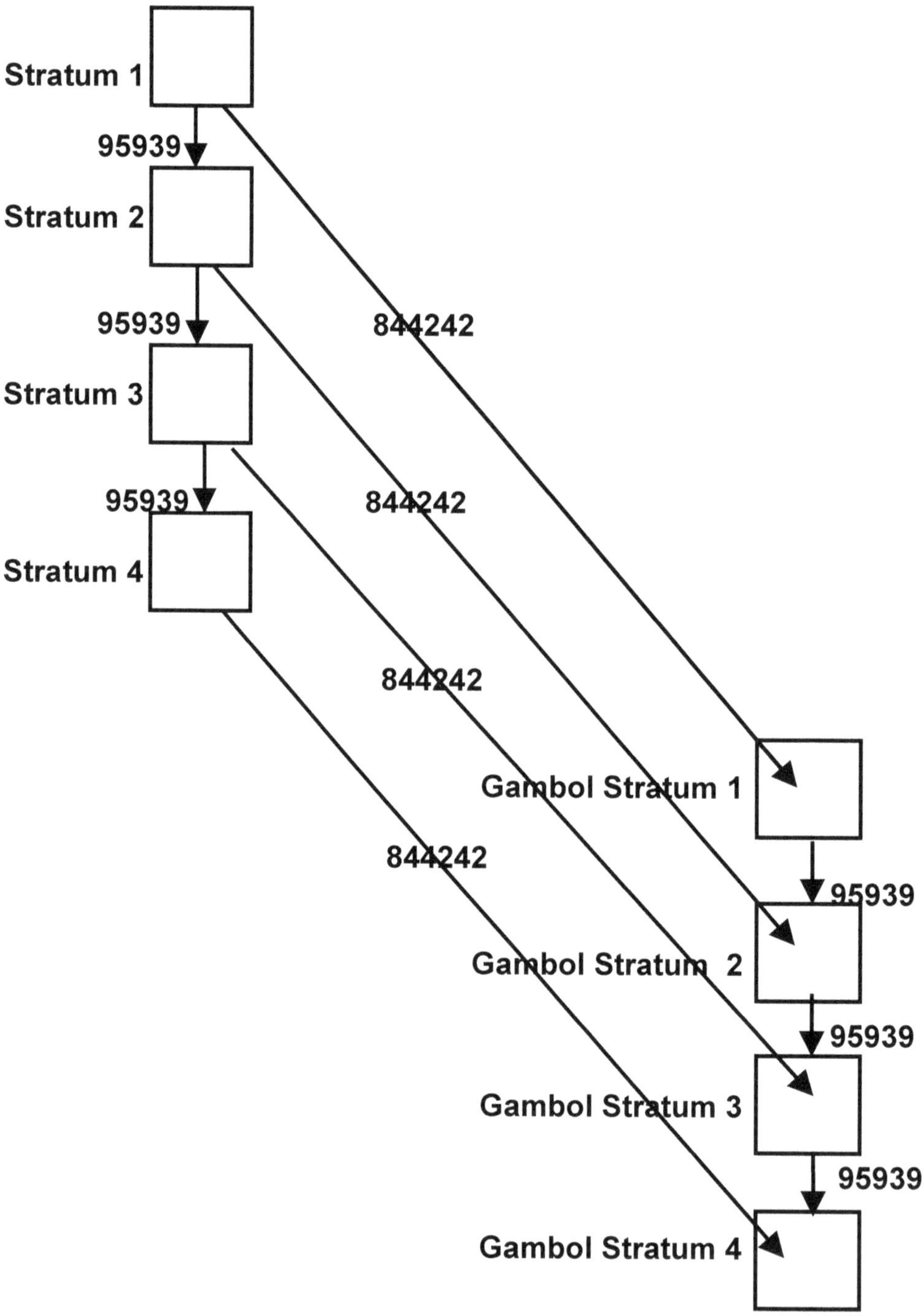

Figure 5.2. Each left block represents a 256 fermion set. Each right block represents a corresponding 224 gambol fermion set since neutrinos do not have gambols. Coresponding fermions in adjacent strata differ in mass by a factor of 95939 in increasing order from stratum 1. Each gambol fermion is a factor of 1/844224 of the corresponding fermion mass. Each gambol mass is 1/95939 of the mass of the gambol immediately above it. Not drawn to scale. All fundamental fermion particles pictured above are fractal.

6. Vector Boson Fractals

6.1 Vector Boson Masses

The masses of gauge fields, massive vector bosons, seem to also be described with a sequence that gives rise to the concept of gauge fields being fractal. This concept is supported by the appearance of gauge fields in interaction terms with fermions:

$$\bar{\psi} g A \psi \tag{6.1}$$

since the interaction coupling constant g is fractal and the fermion wave functions are also fractal by chapter 5.

We will consider a vector boson mass sequence using the constants and relations:

$$\sin^2\theta_W = 0.231 \qquad \sin\theta_W = 0.4806 \tag{6.2}$$
$$\cos\theta_W = 0.8769$$

$$\cos\theta_W = m_{W_3}/m_Z$$
$$\sin\theta_W = m_{W_0}/m_Z$$

$$m_Z = 91.1880$$
$$m_{W_3} = 0.8769 m_Z = 79.165$$
$$m_{W_0} = 0.4806 m_Z = 43.827$$

$$m_{W+} = m_{W-} = 80.3692$$

with vector boson masses expressed in GeV/c^2.

We now approximate the neutral gauge fields Z and W_3 mass average with

$$m_{average} = (m_{W_3} + m_Z)/2 = 85.18 \tag{6.3}$$

based on the assumption that this choice effectively washes out perturbative parts of the W and Z masses. Then

$$m_{average}/m_{W_0} = 1.94 \approx 2 \tag{6.4}$$

or

$$m_{average} \approx 2\, m_{W_0}$$

We now consider the H^0 gauge field with a mass $m_{H_0} = 125.26$. Its mass satisfies

$$m_{H_0} \approx 2.86\, m_{W_0} \approx 3 m_W \tag{6.5}$$

The sequence that seems to be emerging has the recurrence relation is:[23]

$$m_n = n\, m_{W_0} \tag{6.6}$$

Note that there are many SU(2)⊗U(1) groups and their vector bosons in the UST. See Fig. 6.1.

6.2 Fractal Vector Boson Internals

Eq. 6.6 is similar to the previous sequences – particularly for coupling constants. We therefore express it as a fractal:

$$D(2^{m_n}) = D(2^{m_{W_0}}) \text{ “•” } D(2^n) = n\, m_{W_0} \tag{6.7}$$

If we compare eq. 6.7 to eq. 6.9 below (and in chapter 3) expressed for coupling constants g_i then

$$D(2^{g_0}) \text{ “•” } D(2^k) = D(2^{g_i}) \tag{6.8}$$

where $k = 2^i$ and find

$$g_i = 2^i g_0 \tag{6.9}$$

Comparing eqs. 6.6 and 6.9 we see gauge field masses grow linearly while coupling constant values grow as powers of 2. The growth factor of masses appears to be “logarithmic” relative to the growth factor of coupling constants – an observation also suggestive of fractal relations. The fractal factors in eq. 6.1 further support the fractal interpretation of vector boson masses.

Since there are 8 sets of SU(2)⊗U(1) vector bosons in the UST, there are eight sets of vector bosons that presumably follow the mass sequences of eq. 6.6 and 6.7.

---

[23] We note the $m_{average}$ value is approximately equal to ½ of the top quark mass of 172.69. In addition the $m_4 = 175.3$ value in the eq. 6.6 sequence is approximately equal to the top quark mass within 2%.

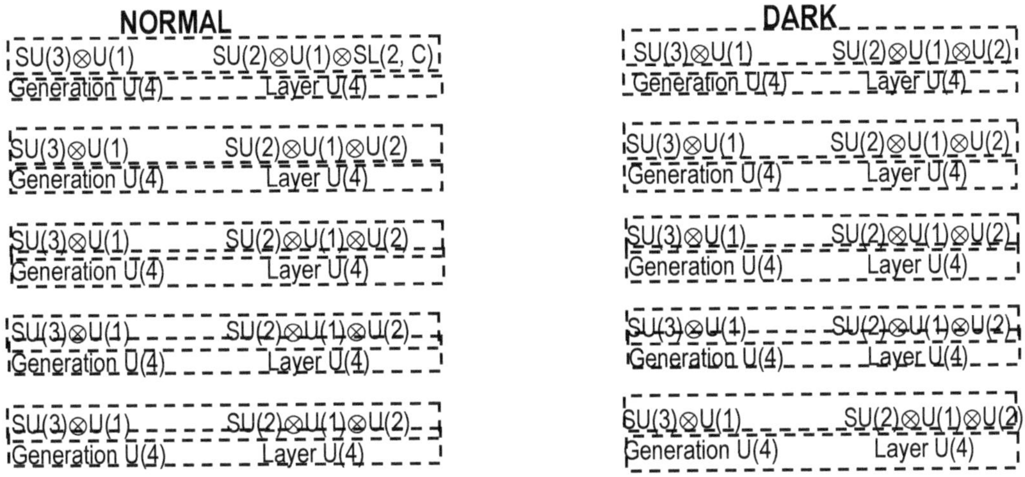

Figure 6.1. The sets of symmetries in the UST). Note each element has a 16 real dimension representation as in the SU(2) dimension array described in eq. 6.8. This depiction appears in the UST. The SL(2, **C**) representation has four dimensions.[24]

[24] The Lorentz Group $SO^+(1, 3)$ is often specified with an SL(2, **C**) representation.

7. Universe Fractals

A universe grows as a one dimension fractal curve that is inverse[25] mapped to a Z-order Lebesgue r dimension space filling curve (Morton space-filling curve). This r dimension curve can then be made into a filled r dimension hypercube. The hypercube can be scaled to be an r dimension space-time within which dynamics can occur.

A universe is generated dynamically – perhaps in a Parent universe. The generation process creates a fractal curve for the universe which is "instantly" mapped to an r dimension space with a corresponding dimension array that structures the particles and symmetries.

Given the total energy (measured in its Parent universe) and its dimension array specifying its fundamental particles and interactions the universe proceeds to evolve dynamically.

A universe is a fractal construct embodying complex dynamics.

[25] A Z-order Lebesgue curve (Morton space-filling curve) can map multi-dimension matter to a one dimension curve.

REFERENCES

Akhiezer, N. I., Frink, A. H. (tr), 1962, *The Calculus of Variations* (Blaisdell Publishing, New York, 1962).
Bjorken, J. D., Drell, S. D., 1964, *Relativistic Quantum Mechanics* (McGraw-Hill, New York, 1965).
Bjorken, J. D., Drell, S. D., 1965, *Relativistic Quantum Fields* (McGraw-Hill, New York, 1965).
Blaha, S., 1995, *C++ for Professional Programming* (International Thomson Publishing, Boston, 1995).
_______, 1998, *Cosmos and Consciousness* (Pingree-Hill Publishing, Auburn, NH, 1998 and 2002).
_______, 2002, *A Finite Unified Quantum Field Theory of the Elementary Particle Standard Model and Quantum Gravity Based on New Quantum Dimensions™ & a New Paradigm in the Calculus of Variations* (Pingree-Hill Publishing, Auburn, NH, 2002).
_______, 2004, *Quantum Big Bang Cosmology: Complex Space-time General Relativity, Quantum Coordinates™ Dodecahedral Universe, Inflation, and New Spin 0, ½, 1 & 2 Tachyons & Imagyons* (Pingree-Hill Publishing, Auburn, NH, 2004).
______, *2005a, Quantum Theory of the Third Kind: A New Type of Divergence-free Quantum Field Theory Supporting a Unified Standard Model of Elementary Particles and Quantum Gravity based on a New Method in the Calculus of Variations* (Pingree-Hill Publishing, Auburn, NH, 2005).
______, 2005b, *The Metatheory of Physics Theories, and the Theory of Everything as a Quantum Computer Language* (Pingree-Hill Publishing, Auburn, NH, 2005).
______, 2005c, *The Equivalence of Elementary Particle Theories and Computer Languages: Quantum Computers, Turing Machines, Standard Model, Superstring Theory, and a Proof that Gödel's Theorem Implies Nature Must Be Quantum* (Pingree-Hill Publishing, Auburn, NH, 2005).
______, 2006a, *The Foundation of the Forces of Nature* (Pingree-Hill Publishing, Auburn, NH, 2006).
______, 2006b, *A Derivation of ElectroWeak Theory based on an Extension of Special Relativity; Black Hole Tachyons; & Tachyons of Any Spin.* (Pingree-Hill Publishing, Auburn, NH, 2006).
_______, 2007a, *Physics Beyond the Light Barrier: The Source of Parity Violation, Tachyons, and A Derivation of Standard Model Features* (Pingree-Hill Publishing, Auburn, NH, 2007).
______, 2007b, *The Origin of the Standard Model: The Genesis of Four Quark and Lepton Species, Parity Violation, the ElectroWeak Sector, Color SU(3), Three Visible Generations of Fermions, and One Generation of Dark Matter with Dark Energy* (Pingree-Hill Publishing, Auburn, NH, 2007).
______, *2008a, A Direct Derivation of the Form of the Standard Model From GL(16) (Pingree-Hill Publishing, Auburn, NH, 2008).*
______, 2008b, *A Complete Derivation of the Form of the Standard Model With a New Method to Generate Particle Masses Second Edition* (Pingree-Hill Publishing, Auburn, NH, 2008)
______, 2009, *The Algebra of Thought & Reality: The Mathematical Basis for Plato's Theory of Ideas, and Reality Extended to Include A Priori Observers and Space-Time Second Edition* (Pingree-Hill Publishing, Auburn, NH, 2009).
______, 2009a, *Bright Stars, Bright Universe* (Pingree-Hill Publishing, Auburn, NH, 2008)
______, 2010a, *Operator Metaphysics: A New Metaphysics Based on a New Operator Logic and a New Quantum Operator Logic that Lead to a Mathematical Basis for Plato's Theory of Ideas and Reality* (Pingree-Hill Publishing, Auburn, NH, 2010).
______, 2010b, *The Standard Model's Form Derived from Operator Logic, Superluminal Transformations and GL(16)* (Pingree-Hill Publishing, Auburn, NH, 2010).
______, 2010c, *SuperCivilizations: Civilizations as Superorganisms* (McMann-Fisher Publishing, Auburn, NH, 2010).
______, 2011a, *21st Century Natural Philosophy Of Ultimate Physical Reality* (McMann-Fisher Publishing, Auburn, NH, 2011).
______, 2011b, *All the Universe! Faster Than Light Tachyon Quark Starships & Particle Accelerators with the LHC as a Prototype Starship Drive Scientific Edition* (Pingree-Hill Publishing, Auburn, NH, 2011).
______, 2011c, *From Asynchronous Logic to The Standard Model to Superflight to the Stars* (Blaha Research, Auburn, NH, 2011).

______, 2012a, *From Asynchronous Logic to The Standard Model to Superflight to the Stars volume 2: Superluminal CP and CPT, U(4) Complex General Relativity and The Standard Model, Complex Vierbein General Relativity, Kinetic Theory, Thermodynamics* (Blaha Research, Auburn, NH, 2012).
______, 2012b, *Standard Model Symmetries, And Four And Sixteen Dimension Complex Relativity; The Origin Of Higgs Mass Terms* (Blaha Reasearch, Auburn, NH, 2012).
______, 2013a, *Multi-Stage Space Guns, Micro-Pulse Nuclear Rockets, and Faster-Than-Light Quark-Gluon Ion Drive Starships* (Blaha Research, Auburn, NH, 2013).
______, 2013b, *The Bridge to Dark Matter; A New Sibling Universe; Dark Energy; Inflatons; Quantum Big Bang; Superluminal Physics; An Extended Standard Model Based on Geometry* (Blaha Reasearch, Auburn, NH, 2013).
______, 2014a, *Universes and Megaverses: From a New Standard Model to a Physical Megaverse; The Big Bang; Our Sibling Universe's Wormhole; Origin of the Cosmological Constant, Spatial Asymmetry of the Universe, and its Web of Galaxies; A Baryonic Field between Universes and Particles; Megaverse Extended Wheeler-DeWitt Equation* (Blaha Reasearch, Auburn, NH, 2014).
______, 2014b, *All the Megaverse! Starships Exploring the Endless Universes of the Cosmos Using the Baryonic Force* (Blaha Research, Auburn, NH, 2014).
______, 2014c, *All the Megaverse! II Between Megaverse Universes: Quantum Entanglement Explained by the Megaverse Coherent Baryonic Radiation Devices – PHASERs Neutron Star Megaverse Slingshot Dynamics Spiritual and UFO Events, and the Megaverse Microscopic Entry into the Megaverse* (Blaha Research, Auburn, NH, 2014).
______, 2015a, *PHYSICS IS LOGIC PAINTED ON THE VOID: Origin of Bare Masses and The Standard Model in Logic, U(4) Origin of the Generations, Normal and Dark Baryonic Forces, Dark Matter, Dark Energy, The Big Bang, Complex General Relativity, A Megaverse of Universe Particles* (Blaha Research, Auburn, NH, 2015).
______, 2015b, *PHYSICS IS LOGIC Part II: The Theory of Everything, The Megaverse Theory of Everything, U(4)⊗U(4) Grand Unified Theory (GUT), Inertial Mass = Gravitational Mass, Unified Extended Standard Model and a New Complex General Relativity with Higgs Particles, Generation Group Higgs Particles* (Blaha Research, Auburn, NH, 2015).
____, 2015c, *The Origin of Higgs ("God") Particles and the Higgs Mechanism: Physics is Logic III, Beyond Higgs – A Revamped Theory With a Local Arrow of Time, The Theory of Everything Enhanced, Why Inertial Frames are Special, Universes of the Mind* (Blaha Research, Auburn, NH, 2015).
______, 2015d, *The Origin of the Eight Coupling Constants of The Theory of Everything: U(8) Grand Unified Theory of Everything (GUTE), S^8 Coupling Constant Symmetry, Space-Time Dependent Coupling Constants, Big Bang Vacuum Coupling Constants, Physics is Logic IV* (Blaha Research, Auburn, NH, 2015).
______, 2016a, *New Types of Dark Matter, Big Bang Equipartition, and A New U(4) Symmetry in the Theory of Everything: Equipartition Principle for Fermions, Matter is 83.33% Dark, Penetrating the Veil of the Big Bang, Explicit QFT Quark Confinement and Charmonium, Physics is Logic V* (Blaha Research, Auburn, NH, 2016).
______, 2016b, *The Periodic Table of the 192 Quarks and Leptons in The Theory of Everything: The U(4) Layer Group, Physics is Logic VI* (Blaha Research, Auburn, NH, 2016).
______, 2016c, *New Boson Quantum Field Theory, Dark Matter Dynamics, Dark Matter Fermion Layer Mixing, Genesis of Higgs Particles, New Layer Higgs Masses, Higgs Coupling Constants, Non-Abelian Higgs Gauge Fields, Physics is Logic VII* (Blaha Research, Auburn, NH, 2016).
______, 2016d, *Unification of the Strong Interactions and Gravitation: Quark Confinement Linked to Modified Short-Distance Gravity; Physics is Logic VIII* (Blaha Research, Auburn, NH, 2016).
______, 2016e, *MoND: Unification of the Strong Interactions and Gravitation II, Quark Confinement Linked to Large-Scale Gravity, Physics is Logic IX* (Blaha Research, Auburn, NH, 2016).
______, 2016f, *CQ Mechanics: A Unification of Quantum & Classical Mechanics, Quantum/Semi-Classical Entanglement, Quantum/Classical Path Integrals, Quantum/Classical Chaos* (Blaha Research, Auburn, NH, 2016).
______, 2016g, *GEMS Unified Gravity, ElectroMagnetic and Strong Interactions: Manifest Quark Confinement, A Solution for the Proton Spin Puzzle, Modified Gravity on the Galactic Scale* (Pingree Hill Publishing, Auburn, NH, 2016).
______, 2016h, *Unification of the Seven Boson Interactions based on the Riemann-Christoffel Curvature Tensor* (Pingree Hill Publishing, Auburn, NH, 2016).
______, 2017a, *Unification of the Eleven Boson Interactions based on 'Rotations of Interactions'* (Pingree Hill Publishing, Auburn, NH, 2017).
______, 2017b, *The Origin of Fermions and Bosons, and Their Unification* (Pingree Hill Publishing, Auburn, NH, 2017).
______, 2017c, *Megaverse: The Universe of Universes* (Pingree Hill Publishing, Auburn, NH, 2017).
______, 2017d, *SuperSymmetry and the Unified SuperStandard Model* (Pingree Hill Publishing, Auburn, NH, 2017).

______, 2017e, *From Qubits to the Unified SuperStandard Model with Embedded SuperStrings: A Derivation* (Pingree Hill Publishing, Auburn, NH, 2017).
______, 2017f, *The Unified SuperStandard Model in Our Universe and the Megaverse: Quarks, ... ,* (Pingree Hill Publishing, Auburn, NH, 2017).
______, 2018a, *The Unified SuperStandard Model and the Megaverse SECOND EDITION A Deeper Theory based on a New Particle Functional Space that Explicates Quantum Entanglement Spookiness (Volume 1)* (Pingree Hill Publishing, Auburn, NH, 2018).
______, 2018b, *Cosmos Creation: The Unified SuperStandard Model, Volume 2, SECOND EDITION* (Pingree Hill Publishing, Auburn, NH, 2018).
______, 2018c, *God Theory (*Pingree Hill Publishing, Auburn, NH, 2018).
______, 2018d, *Immortal Eye: God Theory: Second Edition* (Pingree Hill Publishing, Auburn, NH, 2018).
______, 2018e, *Unification of God Theory and Unified SuperStandard Model THIRD EDITION* (Pingree Hill Publishing, Auburn, NH, 2018).
______, 2019a, *Calculation of: QED α = 1/137, and Other Coupling Constants of the Unified SuperStandard Theory* (Pingree Hill Publishing, Auburn, NH, 2019).
______, 2019b, *Coupling Constants of the Unified SuperStandard Theory SECOND EDITION* (Pingree Hill Publishing, Auburn, NH, 2019).
______, 2019c, *New Hybrid Quantum Big_Bang–Megaverse_Driven Universe with a Finite Big Bang and an Increasing Hubble Constant* (Pingree Hill Publishing, Auburn, NH, 2019).
______, 2019d, *The Universe, The Electron and The Vacuum* (Pingree Hill Publishing, Auburn, NH, 2019).
______, 2019e, *Quantum Big Bang – Quantum Vacuum Universes (Particles)* (Pingree Hill Publishing, Auburn, NH, 2019).
______, 2019f, *The Exact QED Calculation of the Fine Structure Constant Implies ALL 4D Universes have the Same Physics/Life Prospects* (Pingree Hill Publishing, Auburn, NH, 2019).
______, 2019g, *Unified SuperStandard Theory and the SuperUniverse Model: The Foundation of Science* (Pingree Hill Publishing, Auburn, NH, 2019).
______, 2020a, *Quaternion Unified SuperStandard Theory (The QUeST) and Megaverse Octonion SuperStandard Theory (MOST)* (Pingree Hill Publishing, Auburn, NH, 2020).
______, 2020b, *United Universes Quaternion Universe - Octonion Megaverse* (Pingree Hill Publishing, Auburn, NH, 2020).
______, 2020c, *Unified SuperStandard Theories for Quaternion Universes & The Octonion Megaverse* (Pingree Hill Publishing, Auburn, NH, 2020).
______, 2020d, *The Essence of Eternity: Quaternion & Octonion SuperStandard Theories* (Pingree Hill Publishing, Auburn, NH, 2020).
______, 2020e, *The Essence of Eternity II* (Pingree Hill Publishing, Auburn, NH, 2020).
______, 2020f, *A Very Conscious Universe* (Pingree Hill Publishing, Auburn, NH, 2020).
______, 2020g, *Hypercomplex Universe* (Pingree Hill Publishing, Auburn, NH, 2020).
______, 2020h, *Beneath the Quaternion Universe* (Pingree Hill Publishing, Auburn, NH, 2020).
______, 2020i, *Why is the Universe Real? From Quaternion & Octonion to Real Coordinates* (Pingree Hill Publishing, Auburn, NH, 2020).
______, 2020j, *The Origin of Universes: of Quaternion Unified SuperStandard Theory (QUeST); and of the Octonion Megaverse (UTMOST)* (Pingree Hill Publishing, Auburn, NH, 2020).
______, 2020k, *The Seven Spaces of Creation: Octonion Cosmology* (Pingree Hill Publishing, Auburn, NH, 2020).
______, 2020l, *From Octonion Cosmology to the Unified SuperStandard Theory of Particles* (Pingree Hill Publishing, Auburn, NH, 2020).
______, 2021a, *Pioneering the Cosmos* (Pingree Hill Publishing, Auburn, NH, 2021).
______, 2021b, *Pioneering the Cosmos II* (Pingree Hill Publishing, Auburn, NH, 2021).
______, 2021c, *Beyond Octonion Cosmology* (Pingree Hill Publishing, Auburn, NH, 2021).
______, 2021d, *Universes are Particles* (Pingree Hill Publishing, Auburn, NH, 2021).
______, 2021e, *Octonion-like dna-based life, Universe expansion is decay, Emerging New Physics* (Pingree Hill Publishing, Auburn, NH, 2021).
______, 2021f, *The Science of Creation New Quantum Field Theory of Spaces* (Pingree Hill Publishing, Auburn, NH, 2021).
______, 2021g, *Quantum Space Theory With Application to Octonion Cosmology & Possibly To Fermionic Condensed Matter* (Pingree Hill Publishing, Auburn, NH, 2021).
______, 2021h, *21st Century Natural Philosophy of Octonion Cosmology , and Predestination, Fate, and Free Will* (Pingree Hill Publishing, Auburn, NH, 2021).

______, 2021i, *Beyond Octonion Cosmology II : Origin of the Quantum; A New Generalized Field Theory (GiFT); A Proof of the Spectrum of Universes; Atoms in Higher Universes* (Pingree Hill Publishing, Auburn, NH, 2021).
______, 2021j, *Integration of General Relativity and Quantum Theory: Octonion Cosmology, GiFT, Creation/Annihilation Spaces CASe, Reduction of Spaces to a Few Fermions and Symmetries in Fundamental Frames* (Pingree Hill Publishing, Auburn, NH, 2021).
______, 2022a, *New View of Octonion Cosmology Based on the Unification of General Relativity and Quantum Theory* (Pingree Hill Publishing, Auburn, NH, 2022).
______, 2022b, *The Dust Beneath Hypercomplex Cosmology* (Pingree Hill Publishing, Auburn, NH, 2022).
______, 2022c, *Passing Through Nature to Eternity: ProtoCosmos, HyperCosmos, Unified SuperStandard Theory* (Pingree Hill Publishing, Auburn, NH, 2022).
______, 2022d, *HyperCosmos Fractionation and Fundamental Reference Frame Based Unification: Particle Inner Space Basis of Parton and Dual Resonance Models* (Pingree Hill Publishing, Auburn, NH, 2022).
______, 2022e, *A New UniDimension ProtoCosmos and SuperString F-Theory Relation to the HyperCosmos* (Pingree Hill Publishing, Auburn, NH, 2022).
______, 2022f, *The Cosmic Panorama: ProtoCosmos, HyperCosmos, Unified SuperStandard Theory (UST) Derivation* (Pingree Hill Publishing, Auburn, NH, 2022).
_____, 2022g, *Ultimate Origin: ProtoCosmos and HyperCosmos* (Pingree Hill Publishing, Auburn, NH, 2022).
______, 2023a, *UltraUnification and the Generation of the Cosmos* (Pingree Hill Publishing, Auburn, NH, 2023).
______, 2023b, *God and and Cosmos Theory* (Pingree Hill Publishing, Auburn, NH, 2023).
______, 2023c, *A New Completely Geometric SU(8) Cosmos Theory; New PseudoFermion Fields; Fibonacci-like Dimension Arrays; Ramsey Number Approximation* (Pingree Hill Publishing, Auburn, NH, 2023).
______, 2023d, *Newton's Apple is Now the Fermion* (Pingree Hill Publishing, Auburn, NH, 2023).
______, 2023e,*Cosmos Theory: The Sub-Particle Gambol Model* (Pingree Hill Publishing, Auburn, NH, 2023).
_____, 2024a, *Cosmos-Universe-Particle-Gambol Theory* (Pingree Hill Publishing, Auburn, NH, 2024).
_____, 2024b, *Fractal Cosmos Theory* (Pingree Hill Publishing, Auburn, NH, 2024).
_____, 2024c, *Fractal Cosmic Curve: Tensor-Based CosmosTheory* (Pingree Hill Publishing, Auburn, NH, 2024).
_____, 2024d, *The Eternal Form of Cosmos Theory* (Pingree Hill Publishing, Auburn, NH, 2024).
_____, 2024e, *The Eternal Form of Cosmos Theory Third Edition* (Pingree Hill Publishing, Auburn, NH, 2024).
_____, 2024f, *Fundamental Constants of Cosmos Theory and The Standard Model* (Pingree Hill Publishing, Auburn, NH, 2024).
_____, 2024g, *Quark, Lepton, W and Z Masses of Cosmos Theory and The Standard Model* (Pingree Hill Publishing, Auburn, NH, 2024).
_____, 2024h, *Geometric Cosmos Geometric Universe* (Pingree Hill Publishing, Auburn, NH, 2024).
_____, 2024i, *Particles and Universes of Cosmos Theory* (Pingree Hill Publishing, Auburn, NH, 2024).
_____, 2024j, *Unification of the Subluminal and the Superluminal in Cosmos Theory* (Pingree Hill Publishing, Auburn, NH, 2024).
_____, 2024k, *The Dawn of Dynamic Cosmos Dimension Arrays* (Pingree Hill Publishing, Auburn, NH, 2024).
_____, 2024ℓ, *Structure and Dynamics of Cosmos Theory and the Unified SuperStandard Theory* (Pingree Hill Publishing, Auburn, NH, 2024).
_____, 2025a, *Black Holes, White Holes, and Superluminal Starship* (Pingree Hill Publishing, Auburn, NH, 2025).
_____, 2025b, *The Sequence Theory Foundation of Cosmos Theory and Gambol Theory* (Pingree Hill Publishing, Auburn, NH, 2025).
____, 2025c, *BEYOND TIME AND SPACE Fermion Strata & Consistency Conditions for Sequences, Cosmos Spaces and Gambols* (Pingree Hill Publishing, Auburn, NH, 2025).
____, 2025d, *The C o s m o s as a Physics Organism* (Blaha Research, Auburn, NH, 2025).
____, 2025e, *The Root of Reality* (Blaha Research, Auburn, NH, 2025).
Eddington, A. S., 1952, *The Mathematical Theory of Relativity* (Cambridge University Press, Cambridge, U.K., 1952).
Fant, Karl M., 2005, *Logically Determined Design: Clockless System Design With NULL Convention Logic* (John Wiley and Sons, Hoboken, NJ, 2005).
Feinberg, G. and Shapiro, R., 1980, *Life Beyond Earth: The Intelligent Earthlings Guide to Life in the Universe* (William Morrow and Company, New York, 1980).
Gelfand, I. M., Fomin, S. V., Silverman, R. A. (tr), 2000, *Calculus of Variations* (Dover Publications, Mineola, NY, 2000).
Giaquinta, M., Modica, G., Souchek, J., 1998, *Cartesian Coordinates in the Calculus of Variations* Volumes I and II (Springer-Verlag, New York, 1998).
Giaquinta, M., Hildebrandt, S., 1996, *Calculus of Variations* Volumes I and II (Springer-Verlag, New York, 1996).

Gradshteyn, I. S. and Ryzhik, I. M., 1965, *Table of Integrals, Series, and Products* (Academic Press, New York, 1965).
Heitler, W., 1954, *The Quantum Theory of Radiation* (Claendon Press, Oxford, UK, 1954).
Huang, Kerson, 1992, *Quarks, Leptons & Gauge Fields 2nd Edition* (World Scientific Publishing Company, Singapore, 1992).
Jost, J., Li-Jost, X., 1998, *Calculus of Variations* (Cambridge University Press, New York, 1998).
Kaku, Michio, 1993, *Quantum Field Theory*, (Oxford University Press, New York, 1993).
Kirk, G. S. and Raven, J. E., 1962, *The Presocratic Philosophers* (Cambridge University Press, New York, 1962).
Landau, L. D. and Lifshitz, E. M., 1987, *Fluid Mechanics 2nd Edition*, (Pergamon Press, Elmsford, NY, 1987).
Rescher, N., 1967, *The Philosophy of Leibniz* (Prentice-Hall, Englewood Cliffs, NJ, 1967).
Riesz, Frigyes and Sz.-Nagy, Béla, 1990, *Functional Analysis* (Dover Publications, New York, 1990).
Sakurai, J. J., 1964, *Invariance Principles and Elementary Particles* (Princeton University Press, Princeton, NJ, 1964).
Weinberg, S., 1972, *Gravitation and Cosmology* (John Wiley and Sons, New York, 1972).
Weinberg, S., 1995, *The Quantum Theory of Fields Volume I* (Cambridge University Press, New York, 1995).

INDEX

About the Author

Stephen Blaha is a well-known Physicist and Man of Letters with interests in Science, Society and civilization, the Arts, and Technology. He had an Alfred P. Sloan Foundation scholarship in college. He received his Ph.D. in Physics from Rockefeller University. He has served on the faculties of several major universities. He was also a Member of the Technical Staff at Bell Laboratories, a manager at the Boston Globe Newspaper, a Director at Wang Laboratories, and President of Blaha Software Inc. and of Janus Associates Inc. (NH).

Among other achievements he was a co-discoverer of the "r potential" for heavy quark binding developing the first (and still the only demonstrable) non-Aeolian gauge theory with an "r" potential; first suggested the existence of topological structures in superfluid He-3; first proposed Yang-Mills theories would appear in condensed matter phenomena with non-scalar order parameters; first developed a grammar-based formalism for quantum computers and applied it to elementary particle theories; first developed a new form of quantum field theory without divergences (thus solving a major 60 year old problem that enabled a unified theory of the Standard Model and Quantum Gravity without divergences to be developed); first developed a formulation of complex General Relativity based on analytic continuation from real space-time; first developed a generalized non-homogeneous Robertson-Walker metric that enabled a quantum theory of the Big Bang to be developed without singularities at t = 0; first generalized Cauchy's theorem and Gauss' theorem to complex, curved multi-dimensional spaces; received Honorable Mention in the Gravity Research Foundation Essay Competition in 1978; first developed a physically acceptable theory of faster-than-light particles; first derived a composition of extremums method in the Calculus of Variations; first quantitatively suggested that inflationary periods in the history of the universe were not needed; first proved Gödel's Theorem implies Nature must be quantum; provided a new alternative to the Higgs Mechanism, and Higgs particles, to generate masses; first showed how to resolve logical paradoxes including Gödel's Undecidability Theorem by developing Operator Logic and Quantum Operator Logic; first developed a quantitative harmonic oscillator-like model of the life cycle, and interactions, of civilizations; first showed how equations describing superorganisms also apply to civilizations. A recent book shows his theory applies successfully to the past 14 years of history and to *new* archaeological data on Andean and Mayan civilizations as well as Early Anatolian and Egyptian civilizations.

He first developed an axiomatic derivation of the form of The Standard Model from geometry – space-time properties – The Unified SuperStandard Model. It unifies all the known forces of Nature. It also has a Dark Matter sector that includes a Dark ElectroWeak sector with Dark doublets and Dark gauge interactions. It uses quantum coordinates to remove infinities that crop up in most interacting quantum field theories and additionally to remove the infinities that appear in the Big Bang and generate inflationary growth of the universe. It shows gravity has a MOND-like form without sacrificing Newton's Laws. It relates the interactions of the MOND-like sector of gravity with the r-potential of Quark Confinement. The axioms of the theory lead to the question of their origin. We suggest in the preceding edition of this book it can be attributed to an entity with God-like properties. We explore these properties in "God Theory" and show they predict that the Cosmos exists forever although individual universes (or incarnations of our universe) "come and go." Several other important results emerge from God Theory such a functionally triune God. The Unified SuperStandard Theory has many other important parts described in the Current Edition of *The Unified SuperStandard Theory* and expanded in subsequent volumes.

Blaha has had a major impact on a succession of elementary particle theories: his Ph.D. thesis (1970), and papers, showed that quantum field theory calculations to all orders in ladder approximations could not give scaling deep inelastic electron-nucleon scattering. He later showed the eigenvalue equation for the fine structure constant α in Johnson-Baker-Willey QED had a zero at $\alpha = 1$ not 1/137 by solving the Schwinger-Dyson equations to all orders in an approximation that agreed with exact results to 4th order in α thus ending interest in this theory. In 1979 at Prof. Ken Johnson's (MIT) suggestion he calculated the proton-neutron mass difference in the MIT bag model and found the result had the wrong sign reducing interest in the bag model. These results all appear in Physical Review papers. In the 2000's he repeatedly pointed out the shortcomings of SuperString theory and showed that The Standard Model's form could be derived from space-time geometry by an extension of Lorentz transformations to faster than light transformations. This deeper space-time basis greatly increases the possibility that it is part of THE fundamental theory. Recently, Blaha showed that the Weak interactions differed significantly from the Strong, electromagnetic and gravitation interactions in important respects while these interactions had similar features, and suggested that ElectroWeak theory, which is essentially a glued union of the Weak interactions and Electromagnetism, possibly modulo unknown Higgs particle features, be replaced by a unified theory of the other interactions combined with a stand-alone Weak interaction theory. Blaha also showed that, if Charmonium calculations are taken seriously, the Strong interaction coupling constant is only a factor of five larger than the electromagnetic coupling constant, and thus Strong interaction perturbation theory would make sense and yield physically meaningful results.

In graduate school (1965-71) he wrote substantial papers in elementary particles and group theory: The Inelastic E- P Structure Functions in a Gluon Model. Phys. Lett. B40:501-502,1972; Deep-Inelastic E-P Structure Functions In A Ladder Model With Spin 1/2 Nucleons, Phys.Rev. D3:510-523,1971; Continuum Contributions To The Pion Radius, Phys. Rev. 178:2167-2169,1969; Character Analysis of U(N) and SU(N), J. Math. Phys. 10, 2156 (1969); and The Calculation of the Irreducible Characters of the Symmetric Group in Terms of the Compound Characters, (Published as Blaha's Lemma in D. E. Knuth's book: *The Art of Computer Programming Vols. 1 – 4*).

In the early 1980's Blaha was also a pioneer in the development of UNIX for financial, scientific and Internet applications: benchmarked UNIX versions showing that block size was critical for UNIX performance, developing financial modeling software, starting database benchmarking comparison studies, developing Internet-like UNIX networking (1982) and developing a hybrid shell programming technique (1982) that was a precursor to the PERL programming language. He was also the manager of the AT&T ten-year future products development database. His work helped lead to commercial UNIX on computers such as Sun Micros, IBM AIX minis, and Apple computers.

In the 1980's he pioneered the development of PC Desktop Publishing on laser printers and was nominated for three "Awards for Technical Excellence" in 1987 by PC Magazine for PC software products that he designed and developed.

Recently he has developed a theory of Megaverses – actual universes of which our universe is one – with quantum particle-like properties based on the Wheeler-DeWitt equation of Quantum Gravity. He has developed a theory of a baryonic force, which had been conjectured many years ago, and estimated the strength of the force based on discrepancies in measurements of the gravitational constant G. This force, operative in D-dimensional space, can be used to escape from our universe in "uniships" which are the equivalent of the faster-than-light starships proposed in the author's earlier books. Thus travel to other universes, as well as to other stars is possible.

Blaha also considered the complexified Wheeler-DeWitt equation and showed that its limitation to real-valued coordinates and metrics generated a Cosmological Constant in the Einstein equations.

The author has also recently written a series of books on the serious problems of the United States and their solution as well as a book on the decline of Mankind that will follow from current social and genetic trends in Mankind.

In the past twenty years Dr. Blaha has written over 80 books on a wide range of topics. Some recent major works are: *From Asynchronous Logic to The Standard Model to Superflight to the Stars*, *All the Universe!*, *SuperCivilizations: Civilizations as Superorganisms, America's Future: an Islamic Surge, ISIS, al Qaeda, World Epidemics, Ukraine, Russia-China Pact, US Leadership Crisis, The Rises and Falls of Man – Destiny – 3000 AD: New Support for a Superorganism MACRO-THEORY of CIVILIZATIONS From CURRENT WORLD TRENDS and NEW Peruvian, Pre-Mayan, Mayan, Anatolian, and Early Egyptian Data, with a Projection to 3000 AD,* and *Mankind in Decline: Genetic Disasters, Human-Animal Hybrids, Overpopulation, Pollution, Global Warming, Food and Water Shortages, Desertification, Poverty, Rising Violence, Genocide, Epidemics, Wars, Leadership Failure.*

He has taught approximately 4,000 students in undergraduate, graduate, and postgraduate corporate education courses primarily in major universities, and large companies and government agencies.

He developed a quantum theory, The Unified SuperStandard Theory (UST), which describes elementary particles in detail without the difficulties of conventional quantum field theory. He found that the internal symmetries of this theory could be exactly derived from an octonion theory called QUeST. He further found that another octonion theory (UTMOST) describes the Megaverse. It can hold QUeST universes such as our own universe. It has an internal symmetry structure which is a superset of the QUeST internal symmetries.

Recently he developed Octonion Cosmology. He replaced it with HyperCosmos theory, which has significantly better features. He developed a fractionalization process for dimensions, particles and symmetry groups. He also described transformation that reduced particles and dimensions to a far more compact form. He also developed a precursor theory ProtoCosmos that leads to the HyperCosmos.

The author showed that space-time and Internal Symmetries can be unified in any of the ten HyperCosmos spaces in their associated HyperUnification spaces. The combined set of HyperUnification spaces enable all HyperCosmos dimensions to be obtained by a General Relativistic transformation from one primordial dimension in the 42 space-time dimension unified HyperUnification space.

At present the author devel;oped the Cosmos Theory that incorporates ProtoCosmos Theory, HyperCosmos Theory, Limos Theory, Second Kind HyperCosmos Theory and HyperUnification Spaces. He has introduced PseudoFermion wave functions and theory, He has related Cosmos Theory to Regge trajectories of spaces, parton theory, Veneziano amplitudes, Fibonacci numbers and Ramsey numbers. He has calculated an approximation to the difficult R(n,n) Ramsey numbers.

He has developed a Gambol Model that successfully accounts for e-p deep inelastic scattering, fundamental particle resonances, hadron scattering, and the inner structure of particles based on confinement through Casimir forces of ideal gambol gases. The Gambol Planckian Distribution was derived.

He has applied the Gambol Model to particles, universes, and the Cosmos of universes. He showed that the Cosmos may have a distribution of 23 universes corresponding to various Cosmos spaces.

Recently he showed that Cosmos Theory follows from the number of independent asymmetric tensors in a dimension r. He also showed the close parallel between the form of γ-matrices and Cosmos Theory dimension arrays. The closeness suggested that dimension arrays have the same importance as γ-matrices for fermions.

He demonstrated that the pressure of fermions within a space of dimension r balances the Casimir vacuum energy force for 18 dimensions. He showed that $2e\pi = 17.02$ marks the critical point where pressure balances Casimir force, which implies $r = 18$ is the highest dimension Physical Cosmos space. The dimension $2e\pi$ appears to set the approximate dimension for Cosmos spaces with dimension array size $2^{r+4} \cong (17.02/8)^{r+4}. \cong (e\pi/4)^{r+4}. \cong 2.13^{r+4}$.

Now he has found the sequences of Coupling Constant values and fermion masses in the Standard Model and UST. This book unifies the Cosmos spaces spectrum, the coupling constant spectrum and the fermion mass sequences in a fundamental scaling differential equation.

This book shows the fractal basis of Cosmos Theory.

www.ingramcontent.com/pod-product-compliance
Lightning Source LLC
LaVergne TN
LVHW071339210926
848486LV00003B/188